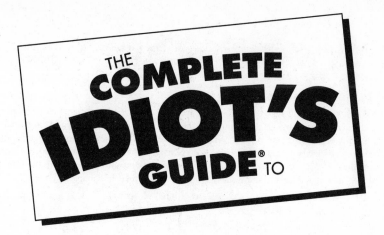

Global Warming

D1301290

by Michael Tennesen

ALPHA

A member of Penguin Group (USA) Inc.

I dedicate this book to my wife, Maggie, and my mom, Ann.
I couldn't have done it without you two.

Copyright © 2004 by Michael Tennesen

THE COMPLETE IDIOT'S GUIDE TO and Design are registered trademarks of Penguin Group (USA) Inc.

International Standard Book Number: 1-59257-071-2
Library of Congress Catalog Card Number: 2003116923

06 05 04   8 7 6 5 4 3 2 1

Interpretation of the printing code: The rightmost number of the first series of numbers is the year of the book's printing; the rightmost number of the second series of numbers is the number of the book's printing. For example, a printing code of 04-1 shows that the first printing occurred in 2004.

Printed in the United States of America

Note: This publication contains the opinions and ideas of its author. It is intended to provide helpful and informative material on the subject matter covered. It is sold with the understanding that the author and publisher are not engaged in rendering professional services in the book. If the reader requires personal assistance or advice, a competent professional should be consulted.

The author and publisher specifically disclaim any responsibility for any liability, loss, or risk, personal or otherwise, which is incurred as a consequence, directly or indirectly, of the use and application of any of the contents of this book.

Most Alpha books are available at special quantity discounts for bulk purchases for sales promotions, premiums, fund-raising, or educational use. Special books, or book excerpts, can also be created to fit specific needs.

For details, write: Special Markets, Alpha Books, 375 Hudson Street, New York, NY 10014.

Publisher: *Marie Butler-Knight*
Product Manager: *Phil Kitchel*
Senior Managing Editor: *Jennifer Chisholm*
Acquisitions Editor: *Mikal Belicove*
Development Editor: *Michael Koch*
Production Editor: *Megan Douglass*
Copy Editor: *Ross Patty*
Illustrator: *Richard King*
Cover/Book Designer: *Trina Wurst*
Indexer: *Tonya Heard*
Layout/Proofreading: *Rebecca Harmon, Donna Martin*

Contents at a Glance

Contents

Appendixes

Foreword

Global warming is an enormously complex—and sometimes controversial—topic, so how can any "idiot" possibly fathom it? Anyone reading *The Complete Idiot's Guide to Global Warming* will soon learn that neither the author nor the readers of this excellent guide are the "idiots." Rather, it is those who make knee-jerk, black-or-white statements that climate change is either "good for you" or the "end of the world" who are the real idiots, not basing their assertions on the full spectrum of scientific knowledge and understanding. Unfortunately, too many of those abound, and they especially need this concise, credible, and accessible book.

Global warming is a hugely complex topic which is often either incompletely understood or completely misunderstood—sadly, sometimes on purpose by special interests for whom "climate denial" is convenient to their ideology or their clients' interests. This book is an excellent guide to making sense of that complexity, effectively presenting cutting-edge scientific knowledge on the subject in easy-to-read chapters. Interspersed photographs, anecdotes, and informational capsules keep the discussion fresh and conversational—this is no dry textbook! If you are serious about learning about this critically important topic that will only increase in significance in the future, read on. And if you want even more technical details on many of these topics, go to one of our websites (http://stephenschneider.stanford.edu).

Much of the debate surrounding global warming stems from the inherent uncertainties. Even the term "global warming" is not a completely accurate representation of the phenomenon. Scientists now use "global climate change" because we now know that as global warming exerts its influence on the world, not all locations will warm up by the same amount, and some may even get colder. But overall, without question, many more places will get warmer than will get colder—and already have. What real scientists do and are doing regarding global warming is to separate what is well understood from what is still uncertain, and describe the current confidence they have in what they do know about the uncertain aspects. Despite these uncertainties there are several important certainties, as this guide makes clear. The world is already warming, and our actions are the largest cause of the most recent (last few decades) warming trend. If we do not make efforts now and in the future to reduce global warming, it is likely we will face a significant number of extremely unpleasant and possibly even dangerous consequences. That we have some time to fix the problem is the good news in this gloomy projection.

One of the largest sources of confusion about global warming, and the reason a guide like this one is so important, is an issue we call *mediarology*. In reporting political, legal, or other advocacy-dominated stories, it is both natural and appropriate for honest journalists to report "both sides" of an issue. Got the Democrat? Better get the Republican! In science, it's different. There are rarely just two polar opposite sides, but rather a range of potential outcomes. Media or political debates on global warming often attempt to divide all viewpoints into two categories: "run in fear" or "it will all be okay," when in fact these are probably the two least likely outcomes. And when reports present arguments for and against the existence of global warming, they are often giving equal voice to a mainstream, well-established consensus and the opposing views of a few extremists. To the uninformed, each position seems equally credible—after all, each claimant has a Ph.D. But you, the reader of this guide, will quickly see that these falsely dichotomous debates do not pit equally credible sides against each other. That is, although some "experts" and many politicians—especially in the United States, those affiliated with the fossil fuel industry or big car companies—assert that human-induced emissions have little impact on the climate, the vast bulk of those truly knowledgeable about climate change and its potential impacts are convinced there is overwhelming evidence for global warming and legitimate reason for concern. With this book you will have the facts to understand these concerns, and to better decide for yourselves who the real "idiots" are.

Stephen H. Schneider and Michael D. Mastrandrea

Stephen Schneider is Professor of Biological Sciences and the Co-Director of the Center for Environmental Science and Policy at Stanford University. He is the founder and editor of the journal *Climatic Change* and was editor-in-chief of the *Encyclopedia of Climate and Weather*. Schneider has consulted with the administrations of all presidents since Carter on climate change issues and is active in the Intergovernmental Panel on Climate Change. He has authored over 450 scientific papers, books, reviews, and other works on environmental science and policy issues.

Michael D. Mastrandrea is a doctoral candidate in the Interdisciplinary Graduate Program in Environment and Resources at Stanford University, and a Department of Energy Global Change Education Program Fellow. His research includes integrated assessment modeling of the climate and economy as a tool for international and domestic climate policy analysis, forecasting the impacts of the El Niño-Southern Oscillation (ENSO) cycle on agriculture and other sectors as a policymaking tool for affected countries, and analyzing the effects of global climate change on world ecosystems.

Introduction

A friend of mine called me the other day. Like most of us, my friend is a concerned citizen. But he wasn't into any long explanation. His questions were pointed: What is global warming? What is it going to do? Is it going to affect me in my lifetime?

I tried to be succinct. Global warming is the effect that the greenhouse gases (carbon dioxide, methane, and others) are having on the earth's climate. Though they constitute less than 1 percent of the atmosphere, the greenhouse gases, especially carbon dioxide, act like a blanket covering the globe. Without the greenhouse effect, the average global temperature would be around 0°F (−18°C). With the greenhouse effect, the temperature averages a toasty 59°F (15°C).

The trouble is the volume of greenhouse gases in the atmosphere has been increasing. The increase is due to the pollution caused by the burning of fossil fuels (coal, oil, and natural gas). It's also caused by the fact that we're cutting down the rainforest, which used to suck up a lot of that carbon dioxide. At the current rate of increase, the volume of greenhouse gases in the atmosphere will more than double by the end of the century.

Most scientists agree that the increase of greenhouse gases in the atmosphere is changing the climate. The controversy is over the amount of change, what that change will bring, and whether it's worth the high repair bill.

The Intergovernmental Panel on Climate Change, a joint project of the World Meteorological Organization and the United Nations Environment Program, estimates that if carbon dioxide in the atmosphere continues to rise, the world will warm up by 2.5 to 10.4°F (1.4 to 5.8°C) by the end of the twenty-first century.

If the temperature rises only 2.5°F (1.4°C), we might be able to live with it. If it rises 10.4°F (5.8°C), then there are going to be major problems. An increase in the global average surface temperature of 10.4°F (5.8°C) is as great a change as ended the last ice age. Only instead of happening over about 50 centuries, it will happen in less than one. The increase could cause major floods in some parts of the country, droughts in other areas, drown large sections of Florida, and wreak havoc with nature everywhere.

Some results are expected in the next 30 years. Some are happening right now. In the last century there has been a widespread retreat of mountain glaciers outside the poles. Half the ice in the European Alps has melted. Summer Arctic sea ice is 40 percent thinner. Glacier National Park has already lost more than two thirds of its glaciers. In another 30 years, scientists estimate Glacier National Park will be glacierless.

Four Antarctic ice shelves have collapsed. The most dramatic of these collapses was the Larsen Ice Shelf, which in 1995 disintegrated over the span of five days. At one point a slab of ice the size of Rhode Island shattered free and floated away.

This book attempts to give you a grasp of the enormity of these issues, the politics behind change, and what you can do about it.

What You'll Learn in This Book

The Complete Idiot's Guide to Global Warming is divided into six parts:

Part 1, "Global Warming Signs," goes into the basics of global warming. I start with predictions of what could happen if we forget about it—just to keep you awake. Next I discuss the atmospheric chemicals that make up the greenhouse effect. Then I'll take you on a four-billion-year tour of the atmosphere. I'll close with a look at the ozone hole, our first climate catastrophe.

Part 2, "A Historical Perspective," shows you how scientists are getting a better look at the history of our climate by looking at ocean bottoms, tree rings, and polar ice. I'll discuss how the climate has made changes—*big* changes—in the past and how we might not be ready for the future. I'll talk about how the climate affected the Vikings and the Mayans, how industrial pollution has tinted the air in the twentieth century, and how cutting down the forests has changed the atmosphere we breathe.

Part 3, "A Meteorological Primer," explains some of the basics of weather and climate. I'll tell you how winds are formed, how rain is made, and how hurricanes are born. I'll talk about El Niño—a sample of what a little warming can do. I'll look at the atmospheric variables that could be causing global warming. I'll also take a look at polar ice, deep ocean currents, and some of the other stuff that could make global warming a lot more dangerous.

Part 4, "The Crystal Ball," discusses some of the effects that global warming might have on Africa, Asia, Europe, Latin America, and our low-lying island nations. I'll also gaze into the future of North America. How will global warming affect commerce, agriculture, and health? You'll learn what might happen to our wildlife, plants, and national parks. And I'll take a look at climate models and how they come up with their predictions for the future.

Part 5, "The Politics," looks at the Kyoto treaty, how it was built, what it required, and where it stands now. I'll look at the international debate, who's skeptical and who's not. This part also covers how U.S. politics has evolved on the question, and whether the causes of pollution reduction, energy conservation, and rainforest preservation aren't worthwhile all by themselves. Finally, I'll sum up the argument.

Part 6, "The Solution," goes beyond the controversy into the answer. What can you do around the home and in the garden to reduce energy consumption? What about that car? Are SUVs and cheap gas such a great deal after all? I'll discuss the rainforest, its value, and what you can do to save it. I'll end with a rundown of great new advances in alternative energies, including solar, wind, and fuel cell technologies.

The appendixes contain further resources and readings for you to explore.

Extras

In addition to the text, you'll find some interesting facts in the following boxed notes:

Climatoids

Here you'll see startling facts and amazing trivia about our climate.

Warm Words

Look here for definitions of global warming terms.

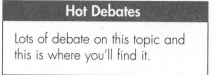

Hot Debates

Lots of debate on this topic and this is where you'll find it.

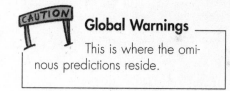

Global Warnings

This is where the ominous predictions reside.

Acknowledgments

I want to thank all the people that helped me assemble the facts for this book. They include: Dr. Richard Alley, Professor of Geosciences, Penn State University; Dorthy Peteet, NASA; Kyle Kudworth, Yerkes Observatory; Lynne Talley, Scripps Institution of Oceanography; Dave Whiticre, Peregrine Fund; Brad Boyle, freelance botanist; Louise Emmons, Smithsonian Insti-tution; Tom Schulenberg, Chicago Field Museum; Richard Somerville, Professor of Meteorology, Scripps Institution of Oceanography; and all the others too numerous to mention that helped me to understand the complexity of this topic.

Also, thanks to my wife Maggie for helping me stay sane while pulling this project together. And thanks to my mother Ann for pointing out those great news stories I missed and for helping me with my spelling.

Special Thanks to the Technical Reviewer

The Complete Idiot's Guide to Global Warming was reviewed by an expert who double-checked the accuracy of what you'll learn here, to help us ensure that this book gives you everything you need to know about global warming. Special thanks are extended to David Reusch.

Trademarks

All terms mentioned in this book that are known to be or are suspected of being trademarks or service marks have been appropriately capitalized. Alpha Books and Penguin Group (USA) Inc. cannot attest to the accuracy of this information. Use of a term in this book should not be regarded as affecting the validity of any trademark or service mark.

Part 1

Global Warming Signs

So what is this thing called "global warming"? It's certainly not a love song. Too many people are arguing over it to have anything to do with love. Global warming is either a catastrophe in the making or it's a great scientific misunderstanding.

In this part, I'll take a look at the basics. I'll run down some of the dire consequences that scientists are predicting if we just ignore the problem. Then I'll take apart the particulars. I'll tell you about the greenhouse effect. I'll get into a little atmospheric chemistry and talk to you about the substances involved. And I'll round out your education with a four-billion-year history of the atmosphere.

I'll also take a look at the ozone hole. How did scientists and politicians get together and tackle this problem, yet seem to be having so much trouble coming up with a consensus on global warming?

Global What?

In This Chapter

- ◆ The inside scoop on rising temperatures
- ◆ How bad is it going to get?
- ◆ What's going to change?
- ◆ Who will be affected?

The idea that the earth is getting warmer may have the scientific community rushing off to conferences all excited, but it's left the general public home sitting on the couch wondering what the fuss is all about. So what if the world gets a few degrees warmer? What are a few degrees? Another sweater? Or better yet, another sweater you don't have to buy. To get you primed for the scientific journey to follow, I'll introduce some of the dire predictions scientists are making if the recent manmade increases in global temperatures—the condition commonly known as global warming—continues.

Who Turned Up the Heat?

Is it really getting warmer? That's what the Intergovernmental Panel on Climate Change (IPCC) says. Established in 1988 by the World Meteorological Organization and the United Nations Environmental Program, the

IPCC consists of a group of over 2,000 scientists. The IPCC concluded in their 2001 report that globally averaged surface temperatures have increased by 1.1°F (.6°C) over the twentieth century. And they also project that by 2100 it will get 2.5 to 10.4°F (1.4 to 5.8°C) hotter than present. The National Research Council and the National Ocean and Atmospheric Administration agree it's getting warmer, and you can look in the mirror if you want to know who's the villain.

Warm Words

The **greenhouse effect** is a natural phenomenon produced by a blanket of gases—carbon dioxide (CO_2), methane (CH_4), nitrous oxide (N_2O), water vapor (H_2O) and others—that act to hold in the reflected heat of the sun and warm up the lower atmosphere.

Global Warnings

If you think the twentieth century was hot, wait until the twenty-first century. The globally averaged surface temperature of the earth is projected to rise 2.5° to 10.4°F (1.4° to 5.8°C) from 1990 to 2100 if manmade greenhouse gas emissions are not reduced. That's a change nearly as dramatic as the one that ended the last ice age.

Scientists report that the 1990s were the warmest decade and 1998 the warmest year since 1861—that's when we started taking the global temperature. However, scientists have used some neat tricks to put together global temperature estimates from tree rings, corals, ice cores, and sediments (discussed in Chapter 5) and believe that the warming of the twentieth century was much greater than any of the previous nine centuries.

What's the cause? Mostly it's our emission of carbon dioxide (CO_2) into the atmosphere, which produces the famous *greenhouse effect*, whereby carbon dioxide and a few other gases help retain the sun's heat. (For more details, see Chapter 2.) About three fourths of the manmade emissions of carbon dioxide to the atmosphere in the last 20 years are due to the combustion of coal, oil, and natural gas. The rest is predominantly due to tilling the soil and cutting down the trees, which removes the vegetation that used to take up the carbon dioxide.

So what's that mean? It means that man is becoming a geological force, comparable to volcanoes and asteroids, both of which have drastically altered the atmosphere in ages past.

Some Catastrophic Results

There have already been some changes in the twentieth century. There has been a widespread retreat of mountain glaciers outside the polar regions. There's been a 40 percent decline in late summer Arctic sea-ice thickness. The *permafrost* is showing signs of thawing. Cherry blossoms are coming out in Washington, D.C. about 4.5 days earlier than they were just 30 years ago. Plant and animal ranges are changing, and some are declining. But the big stuff is projected for the twenty-first century and beyond.

In general, increasing temperatures will put more water into the atmosphere—about 7 percent more for every 1°C (1.8°F) in temperature, due to increased evaporation of the oceans. That's going to cause a lot more rain globally. But it will be uneven—in some places floods, in others droughts. Southeastern forests will be inundated with rain while the plains will change to desert and the desert will get even drier.

Weather will tend to go to the extremes. If the world gets as warm as the IPCC predicts, there could be widespread catastrophe.

> **Warm Words**
>
> The **permafrost** is the vast stretch of permanently frozen subsoil that stretches across the extreme northern latitudes of North America, Europe, and Asia. This land, too cold for the growth of trees, generally marks the northern limits of the forest.

Hurricanes

Scientists predict that the intensity of *hurricane* winds and rain might increase. That means these tremendous tropical storms will get wetter and nastier. Areas like Galveston, Texas, would increasingly be threatened by *storm surges*, massive amounts of seawater pushed ahead by the intense winds of the hurricane. Rising sea levels, from 4 inches to 3 feet (.09 to .88 meters), created worldwide by melting glaciers and warmer expanding seas will also increase storm surges. Storm surges of up to 17 feet created most of the damage and deaths that arose from the Galveston Hurricane of 1900, which took the lives of over 6,000 people. In terms of the loss of life, it was the greatest natural disaster in American history.

> **Warm Words**
>
> **Hurricanes** are violent storms that originate in the tropics and whose winds exceed 73 mph. **Storm surges** are the tremendous rises in tide that storms like these push ahead of them.

And don't forget Hurricane Andrew. When Andrew came ashore in the wee morning hours on Monday, August 21, 1992, it had sustained winds of 145 mph, with gusts approaching 200 mph. Andrew marched through Dade County ripping off shutters, bursting through the windows, tearing off roofs, and tossing mobile homes into the air. It created a zone of devastation in south Dade County that was larger than the city of Chicago. More than 80,000 dwellings were demolished or damaged too severely to live in. Estimates for the total destruction ranged up to $40 billion, the costliest hurricane in U.S. history. With more homes located along shorelines, scientists predict that hurricanes are going to get costlier.

Hurricane winds and rain will increase.

(Photo courtesy of FEMA)

But Andrew and Galveston were rare events, bracketing the two ends of the last 100 years. However, with sea levels rising, rains getting worse, and winds growing stronger, instead of seeing storms of this intensity once a century, they could occur as often as once a decade.

Global Warnings

With the intensity of rain inside hurricanes likely to increase with global warming, storms like Hurricane Mitch may occur more often. Mitch struck Central America in 1998 and killed more than 11,000 people, left 2 million homeless, and caused more than $10 billion in damage across Nicaragua, Honduras, Guatemala, Belize, and southern Mexico. But neither storm surge nor hurricane force winds produced the greatest damage; what took so many lives were the massive mudslides created by these torrential rains.

Forest Fires

Hotter summers, particularly over mid-latitude continental interiors, will increase the risk of drought and ratchet up the chances of forest fires. The year 1988, another record year for heat, produced a scorching summer that seemed to set the world on fire. Large tracts of forest in Yellowstone National Park in the United States and the forests in the countryside of France went up in smoke. The fire in Yellowstone burned 1.2 million acres, about 36 percent of the park. Much of Yellowstone is still charred today.

Hot Debates

Forest fire management is a topic of debate from rural cafés across North America to the Halls of Congress in Washington, D.C. When the Yellowstone fire initially started in early July, 1988, it was allowed to burn. The philosophy then and now in many U.S. national parks is that fire is a part of the natural ecology, and that fire suppression tactics of the past might actually have hurt rather than help the environment. But the Yellowstone fire got out of hand. It burned all through July and August. It wasn't until the first snows in early September that the fire was finally brought under control. With global warming increasing forest fire potential, the role of fire in the natural ecology will be contested further.

The summer of 2002 was a record fire season in the Western United States and Canada as well. A series of fires in Colorado raged through the summer, taking over 400,000 acres. In California, where climate conditions produce a fire season that generally starts in late summer, the drought got brush, trees, and grass burning in spring. In May, I drove by the Chino Hills State Park in southern California—its grassy mountains all lit up in flame. In July, I backpacked into the Sespe Wilderness, 80 miles north of Los Angeles, where fires had closed the roads up until only a few days earlier. On my way home, I watched a fire start on a hillside that erupted with the intensity of gasoline. Cars pulled over, and people were on their cell phones, calling for help.

Floods

Some of the most serious and widespread impacts from global warming will come from flooding, landslides, mudslides, and avalanches driven by sea level rise and increases in rainfall. In the mid and high latitudes of the northern hemisphere, there's already been a two to four percent increase in the frequency of heavy rains over the last 50 years.

Serious and widespread impacts will come from flooding.

(Photo by Michael Tennesen)

Floods in the summer of 2002 raged through Europe. More than 200,000 Czechs were forced to flee their homes in the worst floods in more than a century. The Danube raged too hard for cruise ships, and flooding in former East Germany devastated major towns and tourist centers. German scientists blamed global warming for the deluge.

People that live along rivers are highly susceptible to flooding. Even small rises in water runoff can breach the critical points where dams break, levies burst, and storm drains fill. That point often occurs when the ground reaches its saturation point and can no longer contain water. In areas with lots of groundcover, that takes a while, but in deserts where there's little groundcover to contain the moisture, that point is reached quickly, and runoff can turn into a flash flood.

Climatoids

Often the point where rainfall becomes a destructive flood is not so much a matter of the amount of rain, but how much that rain exceeds the capacity of the land to absorb the moisture. Thus, a season with 30 inches of rainfall in the Southwest United States can create havoc, where such rainfall in the Northeast would be tolerated easily due to denser forest cover and the ability of that cover to absorb moisture.

I witnessed the speed and destruction of a flash flood in Baja California, Mexico, when a storm, only a few hours old, sent a torrent of water down a desert canyon. It skipped over the desert like it was on a greased grill. The force carved out a portion of Baja's Highway 1, upon which we were headed south. My friend and I turned the car around but had to float it across a river further north that hadn't been there a few hours earlier.

Disease

Higher temperature, more rainfall, and more water pooling on the surface is great if you are an insect or insect-borne disease. Mosquito-born malaria could become more common in the United States, as could *West Nile Virus*. Tick-borne Lyme disease might work its way into Canada. Global warming could herald the spread of *encephalitis* and dengue fever.

It will also affect water quality. Higher water temperatures, increased pollutant loads, and the overflow of waste facilities will be bad news for water lovers. Poorer countries, with a lower quality of waste and health facilities to begin with, will suffer the most. Floods will give insect-borne diseases lots of chances to prosper.

In general, infectious diseases will show up more often and hang around longer. As with many of the results of global warming, the rise of infectious disease will be felt the hardest by people in Latin America, Africa, and southeast Asia, especially amongst those countries with the least amount of money to adapt.

In Latin America, insect-borne diseases will expand toward the poles and up the mountains. Extension of the range of infectious disease could take a devastating toll on human health in Africa, already reeling from AIDS. Asia will get its fair share of sick days, too.

Warm Words

West Nile Virus is a disease spread by the bite of an infected mosquito. Though commonly found in Africa, eastern Europe, west Asia, and the Middle East, it didn't occur in the United States until around 1999. Its flulike symptoms are not considered dangerous, though West Nile **encephalitis,** a rare complication of the virus, can be fatal.

System Changes

Global warming is likely to outpace the ability of the natural environment to adapt. Some natural *systems* may undergo significant and irreversible damage. Things like glaciers, coral reefs, atolls, mangroves, boreal and tropical forests, polar and alpine ecosystems, prairie wetlands, and remnant native grasslands will suffer the most. The following subsections introduce some of the larger systems likely to incur the most damage.

Warm Words

Systems from an ecological point of view are niches in the environment that incorporate not just one but various elements such as plants, animals, water, and seasonal weather.

Snowmelt and Glaciers

The snowmelt-dominated watersheds of western North America will experience earlier spring flows and less summer water. The Great Lakes will be lower as will the rivers that flow from them. Areas such as southern California and Morocco will have less rain in the winter—the season that both receive most of their precipitation. Some agricultural areas that rely on melting mountain snows for year-round irrigation, such as California's San Joaquin Valley, will find themselves in trouble. Warmer temperatures will cause more winter moisture to fall as rain, leaving less snow pack to provide water through the summer.

Even in areas with no reduction in rainfall, higher temperatures may create drier soils, and that can stress plants even where there is enough water.

Hot Debates
Water resources are likely to be a hot topic of discussion in arid zones where global warming might lead to a reduction in water supplies. Farmers could see more of their precious irrigation waters routed to growing urban centers rather than to their fields. Cities and farmers along the Euphrates River that flows through Turkey, Syria, and Iraq will experience similar battles as each country promises its people more water than it can deliver. By 2015, according to U.S. and U.N. estimates, as much as 40 percent of the world's population will live in places where there is simply not enough water to go around.

Ice will be rarer in a globally warm world. The ice around the world's poles has been drastically altered, as have most of the world's glaciers. If it keeps getting hotter, within 30 years Glacier National Park may be glacierless. By the end of the twenty-first century half of Europe's alpine glaciers could disappear. The glaciers of Kilimanjaro in Africa and of the Andes in South America are disappearing rapidly.

Coastlines and Estuaries

Global warming will accelerate coastal erosion, reduce wetlands, and increase the salinity of coastal water sources. The latter might alter the water quality in estuaries, interfering with their ability to act as nurseries and stopovers for fish and birds. With rising seas, heavier storms, and lots of floods, more people may opt for the city pool rather than the beach.

Decreased wetlands will impact wildlife.

(Photo by Michael Tennesen)

Rising levels of saltwater could ruin ground waters near tropical and subtropical coastlines as well. Coastal tropical ecosystems such as coral reefs and atolls, salt marshes, and mangrove forests could get thrashed by sea-level rise as well as meaner and more frequent storms.

Coastlines closer to the poles will get their share of woes. Rising sea levels will eat at the shores, less sea ice will produce higher sea levels and bigger waves. A thaw in the permafrost will bring more floods. It won't take too long before everybody is running from the shore.

Forests and Wildlife

Global warming could have drastic effects on the earth's remaining pockets of wilderness, as well. Forests tend to survive in areas that have the right soil, moisture, and temperature. As the world warms up, forests can move poleward as the wind, birds, and animals transport their seeds. But the pace has its limits. Norwegian scientists studying evidence of the movement of forests northward after the last ice age noted that different forest types sometimes took centuries to establish themselves in areas where the climate had changed to one favoring the new forest type.

Climatoids

Plants and animals inside our national parks could find that the ideal climate for growth has shifted poleward. The trouble is plants can't always pick and move to a more hospitable environment. And animals can't follow the habitat poleward because much of the area surrounding our national parks is developed by people.

Increasing carbon dioxide concentrations could actually stimulate the growth of some species. Managed forests—where we get our lumber—will have the easiest time of it, because man can help the forest make adjustments.

Dry forests, in arid or semi-arid areas, where warming is likely to decrease soil moisture, are expected to decrease. Expected droughts in some parts of the Amazon could boost fires already raging to clear the jungle for crops and cattle. Tropical dry forests in Mexico and northern Africa will get even drier.

Climatoids

Ducks, geese, and other waterfowl are likely to be affected if the "prairie pothole" wetlands of the north central United States and southern Canada dry up—as global warming models predict. The prairie potholes are vital areas for breeding, as well as critical stopovers for migration.

Global warming pressures will do a number on wildlife as well. Many species, already at high risk, will be pushed closer to the edge as climate change renders parts of their current habitat unsuitable. Climate pressures in the twenty-first century could cause species currently classified as "critically endangered" to become extinct and the majority of those labeled "endangered or vulnerable" to become rarer and closer to extinction.

Global warming could alter the habitat for grizzly bears, red squirrels, and other wildlife in Yellowstone National Park by reducing the number of whitebark pine trees, an important food source for these critters. Habitat changes in other areas could pressure other species with local or global extinction. Among the IPCC's list of threatened animals are the forest birds in Tanzania, the resplendent quetzal in Central America, the mountain gorilla in Africa, and the Bengal tiger in India.

Global Warnings

Scientists working at the Palmer Station in the Antarctic report that average temperatures there over the last 50 years have risen by 3° to 5°F (1.7° to 2.8°C) in summer and an astronomical 7° to 9°F (3.9° to 5°C) in winter. The heat has ruined much of the dense shrimp-like krill swarms that Adélie penguins feed on in the ocean. Parents now leave hungry chicks alone on the ice while they go out on wide foraging expeditions for diminishing food. Scientists studying the animal warn that the Adélie could be the "canaries in the coal mines" of global warming.

Climate change will lead to loss of habitat for cold and cool water fish, and a gain in habitat for warm water species. Inland lakes and rivers are particularly susceptible to climate change. Global warming could eliminate rainbow trout and brook trout in the

Northeast and upper Midwest. You might have to go to Canada if you want to use your fly pole. Which might not be a bad idea since red fire ants in the Southeast might expand their ranges into the north.

Polar bears, ringed seals, and migratory birds that depend on the Arctic for a living might feel some of the earliest effects since that area is believed to be one of the fastest warming on earth. Already Adélie penguins in the Antarctic are suffering some of the most severe effects of present day global warming.

I'll talk more about the price that wildlife will pay for global warming in Chapter 16.

Agriculture and Industry

In some ways agriculture might benefit from global warming. Increased carbon dioxide levels can stimulate plant growth. However, the optimum climate for certain plants might go north as the climate warms from the equator. It's a varied picture: While Canada might have more rain, the Midwest will be 30 to 60 percent drier.

This could mean that ideal weather for some crops, such as wheat, could move northward into Canada. But these are rapid evolutions in any plant's cycle, and although the weather may be better in Canada, the soils may not be rich enough to sustain the new crops.

One of the most vulnerable industries is the insurance industry. Economic losses from worldwide catastrophic natural events, adjusted for inflation, rose from $3.5 billion in the 1950s to $40 billion in the 1990s—more than a tenfold increase. That's due not only to growing populations, but to increasing development in vulnerable areas, and more frequent storms.

Ocean fish catches might be down, particularly with cold water species such as cod. Expansion of marine aquaculture may partly compensate for those reductions, but aquaculture has its problems with a warmer world as well, since stocks of herring, anchovies, and other fish species used to feed farmed fish might also diminish. Increased sea temperatures may promote algae blooms in fish farms and spread disease.

Tourism will likely suffer. Sea level rise, climbing temperatures, and more rainfall could ruin beaches, the Everglades, and coral reefs off Florida. Global warming will have an equally negative effect on tourism in Latin America. With less ice and less snow, the Swiss Alps may draw a leaner winter crowd.

Climatoids

In general global, warming will affect rural communities more than urban environments. That's because rural communities often depend upon one or two resource-based industries. If your principal industry is corn, and there isn't enough rain, then it's harder to adapt. Whereas in the city people have more options. However, city folk won't go unscathed, because global warming will probably increase the formation of ground-level ozone and other pollutants.

The Bright Side

There are some positive benefits to global warming. Warmer winters in Canada and the northern parts of the United States may significantly reduce energy bills. Plus, the flowers will blossom sooner and birds start nesting earlier—even if the summer is a killer. Because managed forests will do better, lumber should be cheaper. There'll be more water available in water-scarce areas such as Southeast Asia. And tourism may be down in some parts of the world, but perhaps you don't like tourists anyway. Last but not least, fewer people will be freezing to death.

The Net Effect

Basically, the net effect depends on how warm it gets. A few degrees may not do a lot of damage, but the hotter it gets, the more destructive the weather gets. Even for increases of less than a few degrees, the IPCC projects more people will be harmed than benefited by global warming.

Hot Debates

There are certain possibilities called "feedback mechanisms" which some scientists debate could accelerate climate change. Those include significant slowing of the ocean currents that transport warm water to the North Atlantic, major reductions of the Greenland and West Antarctic ice sheets, and large releases of carbon and methane from the land if the permafrost melts. If any of these happen, all bets might be off for the current predicted caps on global warming in the twenty-first century. (For more details see Chapter 13.)

The Least You Need to Know

♦ The average global temperature has risen 1.1°F (.6°C) in the twentieth century.

♦ Scientists predict it will rise 2.5° to 10.4°F (1.4° to 5.8°C) in the twenty-first century.

♦ This might lead to increased damage from hurricanes, floods, fires, and disease.

♦ Major changes will take place in many natural systems, as well as in the economy.

♦ Though there will be some benefits, the net impact will be negative.

Chapter 2

I Don't See Any Greenhouse

In This Chapter

- Man discovers the greenhouse
- The function of the greenhouse
- Name those gases
- Why gases are increasing

If all these ominous weather predictions seem hard to fathom, consider that the causes of all this worry are some simple gases that are present in such small volumes that they are measured in parts per million. But even though their amounts are tiny, their roles are vital to life on Earth. In this chapter you'll learn about the discovery of the greenhouse effect, the gases that make up this chemical blanket, and how we humans are disturbing this phenomenon.

Discovering the Greenhouse Effect

French mathematician Jean-Bapiste-Joseph Fourier first noted the greenhouse effect in 1827 when he compared the atmosphere to a "glass vessel." Fourier was the first to recognize that some of the gases in our atmosphere act like a greenhouse since they let in sunlight—like a greenhouse does—but prevent some of the sun's warmth from radiating back out into space.

In the 1850s, British physicist James Tyndall tried to figure out which of these gasses contributed the most to the warming effect. He measured the heat-trapping ability of various atmospheric gases and was surprised at the results. It seems that the two most common gases, nitrogen and oxygen, which make up 99 percent of the dry atmosphere, have no heat-trapping ability at all. Tyndall concluded that the job of keeping the world warm and toasty fell to carbon dioxide, methane, water vapor, and a few other elements.

Climatoids

The greenhouse gases—carbon dioxide, methane, and others—are long-lived gases (carbon dioxide lasts 100 years in the atmosphere, methane 10 years) that are found in minute quantities within the atmosphere. Water vapor is short-lived and alters their effects. Though the long-lived greenhouse gases play a vital role in keeping the planet warm, they make up less than 1 percent of the total atmosphere. No wonder so many people have such a hard time fathoming the dangers of their imbalance. They are the true phantoms of the atmosphere.

In the 1930s, during a period of particularly hot summers in Europe, George Callendar, a British coal engineer, put together several decades of temperatures taken at weather stations and from shipboard decks and uncovered that the world was gradually getting warmer.

Callendar published his results in 1938. Callendar attributed this to the rising levels of carbon dioxide brought on by the burning of fossil fuels. He was way ahead of his time. However, just after he came out with his results, Europe went into a prolonged cold spell, and nobody seemed to care.

Besides, scientists felt that excess carbon dioxide would be absorbed by the oceans, which acted as vast carbon sinks. Excess carbon dioxide would get sucked in by the seas—and the plants and animals therein—and end up deposited on the sea floor. It wasn't until the 1950s that scientists at the Scripps Institute of Oceanography found that seawater had a limit to its absorption. They projected that as increasing amounts of carbon dioxide were added to the atmosphere, the ocean would reach its saturation point and would no longer act as a ready sponge for man's excesses. Although much of the gas would end up in the sea, at the rate man was adding it to the atmosphere, there would be more than enough left over to influence the climate.

These findings occurred about the time that plans were being made for the International Geophysical Year (July 1957 to December 1958), and the carbon dioxide issue ended up on its agenda. The event was a vast and ambitious project that lasted over 18 months and employed the help of geologists, chemists, climatologists, and other specialists from over 79 nations. It was to be Earth's first physical checkup, and scientists were sent all over the world to measure everything they could.

The Keeling Curve

One young scientist who was sent out on this monumental task was Charles David Keeling. He was fresh off research that had him running around California stuffing samples of the air into bottles that he took back to the lab to measure. Keeling developed a device for measuring carbon dioxide in the atmosphere, the first ever to be able to measure one part per million. He noticed that his samples were 315 parts per million of carbon dioxide, which he would later learn was about 13 percent higher than before the *Industrial Revolution*.

Warm Words

The **Industrial Revolution** is a term applied to the changes that occurred between 1750 and 1850 in Britain when English society transformed from a stable agricultural economy to a modern industrial one. The biggest changes in the United States followed the Civil War from 1860 to 1890. Other countries have had their industrial revolutions in different periods, and some third-world nations are still waiting for the change. From a greenhouse gas perspective, the world's carbon dioxide concentrations started their greatest rise after 1850.

So Keeling decided to expand his research. For the International Geophysical Year, he built two instruments that he called manometers, which were able to take continual readings of the carbon dioxide in the atmosphere. He flew one to Mauna Loa, a high mountain volcano that towered over the big island of Hawaii. He felt that out in the middle of the ocean, high up on a volcano, far from any city and high above the Hawaiian rainforest, he could get what he thought was a pure, undistorted reading of the earth's carbon dioxide content.

In 1958, he began his readings. At first the result was a marvelous record of plant life on our planet. The figure on his chart was a series of humps that represented the various seasons. Carbon dioxide was high in the winter, falling in the spring, low in the summer, and rising again in the fall. What it was recording was the annual global plant cycle. In the increasing warmth of the spring and summer, plants burst into growth, producing leaves, flowers, and fruit, and in that growth process, *photosynthesis* was removing carbon dioxide from the atmosphere. Then in the fall, plant leaves, flowers, and fruit drop to the forest floor. As they disintegrate, the carbon dioxide was released back into the atmosphere.

Keeling was fascinated with these cycles so he kept his machine running way beyond the initial research project. Over several decades his original manometer and a number of others like it were set at various stations around the world to keep track of the atmosphere's volume of carbon dioxide, much like you might keep track of your cholesterol. But when Keeling started collating the data and putting it onto a chart, he uncovered a startling fact. The result is what we now call the Keeling Curve. Although each year had its cyclical bumps, the overall trend was an unmistakable rise upward.

Warm Words

Photosynthesis is the process whereby green plants take in carbon dioxide (CO_2), mostly through their leaves, and water, mostly through their roots, and mix them together in the presence of sunlight and chlorophyll to produce glucose, a carbohydrate utilized by the plant as energy.

Prior to the Industrial Revolution there were approximately 280 parts per million of carbon dioxide. In 1958, there were 315 parts per million. In 1990, there were 355 ppm, and in 2001 there were 370 ppm. That means there is 31 percent more carbon dioxide in the atmosphere now than there was before the Industrial Revolution, and an 18 percent jump in just the last 40 years. In the 1960s Keeling found that the carbon dioxide levels in the atmosphere were rising at 0.7 parts per million per year. In the 1980s the annual rise was up to 1.5 parts per million, more than double.

The trouble was, most of Keeling's work was done prior to 1970, when much of the world was in a cold snap. Though carbon dioxide steadily rose over the course of the century, temperatures in general rose until the 1940s but then dropped back down until the 1970s, when they began rising again toward new records. Ironically, it might have been our own pollution that kept things cool for those middle decades. The National Oceanic and Atmospheric Administration (NOAA) estimates that the surge in coal and oil use during and after World War II sent increasing amounts of particulates into the air which had the effect of cooling the atmosphere down. However, around 1970, the increase in the longer-lived carbon dioxide and methane overwhelmed the shorter-lived sulfates, and the temperature started to rise again.

Hot Debates

We spew about 4 tons of carbon dioxide into the atmosphere for every one of the 6 billion men, women, and children on Earth. Americans contribute closer to 20 tons each. They consume about 22 percent of the world's oil, though they make up only 5 percent of the world's population.

Its Vital Role

The greenhouse effect is not a bad thing. In fact, we need it to stay alive. It helps regulate the temperature of our planet. There are a number of chemical compounds in Earth's atmosphere that make up the greenhouse gases. Greenhouse gases effectively trap in enough of the sun's reflected heat to raise the earth's temperature to what it is today. Without a natural greenhouse effect, the average global temperature of our planet would be around -2°F (-19°C) instead of its present 57°F (14°C).

How'd you like to roast your hot dogs in -0.4°F? It wouldn't be too comfortable. In fact, life as we know it wouldn't exist. So we need the greenhouse effect. No one is arguing that it isn't important. The only disagreement is over whether man is increasing the atmospheric content of these gases and what that will bring.

Greenhouse gases allow direct sunlight, which is relatively short-wave energy, to pass by them unaffected. That's because greenhouse gases are poor absorbers of short-wave energy. It's just not their cup of tea. But as the short-wave energy heats up Earth's surface, it radiates back in the form of longer-wave energy, which is like pizza and ice cream to a greenhouse gas. They gobble up that heat (which would otherwise be lost to space) and warm the lower 10 miles of the atmosphere—where we live, work, and play basketball.

Many of these greenhouse gases occur naturally in the atmosphere, such as carbon dioxide, methane, water vapor, and nitrous oxides. Others, such as chlorofluorocarbons (CFCs), are manmade.

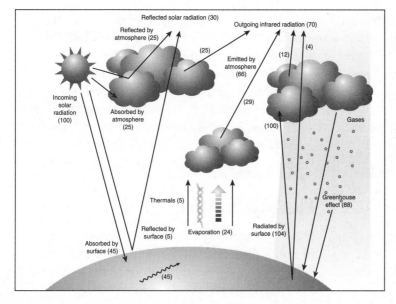

The greenhouse effect is very real.

Carbon Dioxide (CO$_2$)

There are a number of greenhouse gases, but carbon dioxide seems to gather the most attention because it is long lived and causes the greatest amount of global warming of the long-lived gases. We know it from chemistry classes as one of the basic substances of life. We breathe it out and plants take it in. It's the bubbles in your cola, your mineral water, and even your beer. Pretty innocuous if you ask me. How could it be such a problem?

Where It Comes From

Most mammals—including dogs, cats, birds, deer, mountain lions, you, and I—create carbon dioxide when we exhale. Plants take it up and use it in photosynthesis to make leaves, flowers, and fruit. But plants also give off carbon dioxide in fall when those leaves gather on the forest floor. As they disintegrate, the carbon dioxide they took up with photosynthesis is released back into the environment.

Volcanoes also give off carbon dioxide. In fact, this is how the planet got its first atmosphere. The earth's atmosphere was a blanket of carbon dioxide. It was only through the evolution of plant life, which began to take up the carbon dioxide and exhale oxygen, that the planet's atmosphere changed into one that was mostly oxygen. This then allowed for the evolution of animal life. (For more details see Chapter 3.)

Where Is It Stored?

I talked about how plants store carbon dioxide. The atmospheric concentration of carbon dioxide has a marked seasonal shift in the northern hemisphere. That's because the northern hemisphere has more land and more vegetation covering it. Remember Keeling's manometer and how the graph went up and down with the seasons? That seasonal roller coaster is not as pronounced in the southern hemisphere, which has more water. Carbon dioxide is also stored in the oceans, as well as in rocks.

Climatoids

The concentration of carbon dioxide in the atmosphere has increased about 31 percent since 1750, and 18 percent in just the last 40 years. IPCC scientists report that the present amount of carbon dioxide has not been exceeded in the last 420,000 years, and likely not in the last 20 million years.

What's the Problem?

The problem is that carbon dioxide is on the rise in the atmosphere and that's having an effect on the global average temperature. Prior to the Industrial

Revolution, concentrations of carbon dioxide in the atmosphere were relatively stable at about 280 parts per million (ppm). Today they are about 367 ppm and rising, an increase of about 31 percent. They've risen about 18 percent in just the last 40 years. The present carbon dioxide concentration in the atmosphere hasn't been exceeded in the last 420,000 years and probably not in the past 20 million. What makes the problem so unique is that carbon dioxide has a life span in the atmosphere of about 100 years, which means that even if we stopped burning fossil fuels right now, the effects could last for a century.

By the year 2100, IPCC scientists predict that the atmospheric concentrations of carbon dioxide will rise from current levels to a range of from 490 to 1260 ppm (75 to 350 percent above preindustrial figures). About three fourths of these emissions are due to the burning of oil, coal, and natural gas. Some of that gas comes from the more than 500 million cars on the planet. Some of it comes from the fires used to clear the land and create farms and cattle ranches.

Some of it comes from cooking fires. It is estimated that more than two billion people use firewood to fix their daily meals. Open fires are not only inefficient sources of energy but significant contributors to atmospheric carbon dioxide.

CAUTION **Global Warnings**

Since 1900, the average annual consumption of energy has increased more than tenfold. Between 1870 and 1970, the burning of fossil fuels to create this energy added 400 billion tons of carbon dioxide to the atmosphere. From 1970 to 1989 another 400 billion tons were added. The increase is due to the tremendous acceleration in our consumption of coal, oil, and natural gas in the last half of the twentieth century.

We can't just plant trees to get rid of all the extra carbon dioxide. Even if we could put back all the trees we cut down, carbon dioxide would only be reduced by 40 to 70 ppm of the atmosphere. Eventually we are going to have to reduce the amount of carbon dioxide emissions if we are to bring it under control. According to the IPCC, just to stabilize the atmospheric concentration of carbon dioxide in the atmosphere at 450 to 1,000 ppm, global use of coal, oil, and gas will have to be reduced to below 1990 levels and continue to decrease steadily thereafter. Eventually carbon dioxide emissions will have to decline to a small fraction of what they are today in order to stabilize the atmosphere over the long run.

Methane (CH₄)

You might think of methane as the wool blanket of greenhouse gases. That's because it is an extremely efficient absorber of heat. It's 20 to 30 times more efficient at trapping heat than carbon dioxide. Still, methane concentrations in the atmosphere are below 2 parts per million and its lifetime is only about 10 to 12 years. However, that concentration is still about 150 percent higher than it was before the Industrial Revolution.

Methane, the principle component of natural gas.

(Photo by Michael Tennesen)

Methane, also called swamp gas, is the principal component of natural gas. It has both natural and human-influenced sources. Naturally it's given off as part of the decomposition that occurs in wetlands, swamps, and bogs. Man has added a host of other sources to these emissions. They include the bacterial decomposition that occurs in landfills, rice paddies, and the guts of cattle and termites. As man cuts down the forest, termites and bacteria flourish in the refuse left on the forest floor.

> **Climatoids**
>
> As humans continue to expand, so do their cattle. There are over a billion head of cattle on the earth today. Cows produce methane when bacteria break down cellulose in their stomachs. The gas comes out either end of the cow.

Some methane also escapes during the process of coal mining and oil drilling. Methane is also emitted during the incomplete combustion of oil, gas, and coal. Methane is stored in the vast soils of the Arctic tundra, which is why scientists are concerned about the warming of the permafrost. It could release enough methane to drastically affect the climate. There are also huge reserves of methane in the sediments at the bottom of the ocean floor. Because of its efficiency as a greenhouse gas, it wouldn't take much additional release to have an effect on the atmosphere. Methane increases in the atmosphere have slowed in the last 10 years, though scientists are at odds to explain why.

Nitrous Oxide (N_2O)

Like the other greenhouse gases, nitrous oxide has been on the rise since the beginning of the Industrial Revolution. Since nitrogen makes up 78 percent of the atmosphere and oxygen another 21 percent, it's easy to get these two chemicals together. Wherever fuel is burned, it can force nitrogen and oxygen to combine to make nitrous oxide. It's part of those pesky gases that make up photochemical smog, occasionally found in the air over Los Angeles and other major industrial cities.

Microbial processes in soil and water also get these guys going, especially if you toss in some nitrate fertilizers. Nitrate fertilizers have been increasingly used over the last century.

Ozone

There is a broad band of ozone in the stratosphere that I will discuss in Chapter 4. Only small amounts fall out into the lower atmosphere, but in the twentieth century, we've added to that lower atmospheric concentration. Ozone in the lower atmosphere has increased by about 30 percent since the preindustrial era. The IPCC now considers ozone to be the most important greenhouse gas after carbon dioxide and methane. More in Chapter 3.

Chlorofluorocarbons (CFCs)

These gases have no natural sources. Chlorofluorocarbons were synthesized in a laboratory in the late '20s and were initially heralded for their many uses, particularly as refrigerants. But they were also used as aerosol propellants and as cleaning solvents. However, scientists discovered that they gradually accumulated in the environment and were able to destroy ozone in the upper layers of the atmosphere. Thanks to an international agreement, the levels of the major CFCs have peaked and are now declining. (See Chapter 3 for more information.)

Carbon Monoxide (CO) and Others

The carbon monoxide concentration in the atmosphere influences the greenhouse effect because it can modulate the production of methane and ground level ozone. By itself, carbon monoxide does not absorb heat from the sun, as do the other greenhouse gases. The northern hemisphere contains about two times the amount of carbon monoxide as the southern hemisphere partly because people in the north drive a lot more cars.

It's colorless, odorless, and tasteless, but it is harmful to people and animals because carbon monoxide substitutes for oxygen in the blood. The blood is actually more attracted to carbon monoxide than oxygen, although our bodies need oxygen, not carbon monoxide. Carbon monoxide poisoning can begin at concentrations of 20 to 50 ppm. Cigarette smoke contains 400 to 450 ppm. In a busy intersection, concentrations are about 100 ppm. That might be why you have a headache.

Climatoids

Carbon monoxide is one of the gases that may be on the wane. Measured concentrations in the air were generally increasing until the late 1980s when they began to decline. Many scientists believe this is due to the increased usage of catalytic converters on automobiles.

Volatile organic compounds are also greenhouse gases, though their effects are small compared to the others. Their largest source of emissions is from natural vegetation but they are also a byproduct of automobile exhaust and fuel production. Measuring them is extremely difficult. Still, they are thought to be on the increase in the last part of the twentieth century.

Man's contribution to the atmosphere.

(Courtesy IPCC)

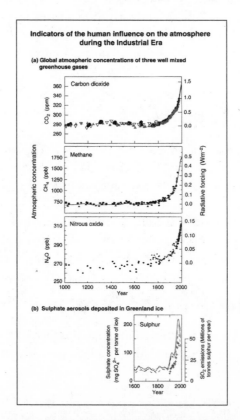

Particulates

Particulates are very small airborne particles or droplets that can influence the climate in a variety of ways. Particulates can absorb the heat radiating off the earth's surface much like greenhouse gases. They also have an effect in clouds, where moisture coalesces on them to create rain. Thus, the life of a particulate in a cloud is rather short, typically no more than a week, which is why particulates are found close to their sources. Unlike carbon dioxide and other greenhouse gases, they don't distribute evenly over the globe.

Particulates are produced in a variety of ways, including the dust from a dust storm, salt from salt water spray, and the ash from volcanic eruptions. They're also in that soot coming from that coal-fired smokestack. And they're in the ash from forest fires. Most particulates are found in the lower atmosphere, but some occur higher up as well. Volcanic eruptions can send particulates in the form of volcanic ash into the upper atmosphere. Several explosive eruptions that have occurred over the last 150 years have had temporary cooling effects on the atmosphere.

Global Warnings

Particulates can make it into the upper atmosphere, either through a volcanic eruption or the rarer impact of an asteroid with the earth. Despite the Hollywoodization of this threat, it's still thought by many scientists that impact of an asteroid with the planet some 65 million years ago might have created such a blanket of dust (particulates) that it blocked out the sun. And this may be why the dinosaurs went extinct.

Water Vapor

Though we saved water for last, it's not because it's the least. Water vapor is actually the most abundant greenhouse gas in the atmosphere. But its effects are more transitory. Perhaps it's better to think of it more as a modulator of the effects of the other longer-lived gases. Water vapor is a principle driving force in the weather. It's also an important feedback mechanism, since as the temperature of the atmosphere rises, more water evaporates from lakes, rivers, the soil, and the sea. The hotter it gets, the more water the atmosphere can hold.

And the more water you have in the air, the more clouds you're likely to have. Clouds can either hold in the heat and make things warm, or they can reflect the sun and cool things down. Scientists using climate models to figure out water vapor's role in a globally warm world claim they still don't have it figured out.

Hot Debates
Water vapor is the most abundant of the greenhouse gases, but scientists still debate its contribution to global warming. As a greenhouse gas, water is an efficient absorber of radiated heat, which can warm the planet significantly. But more water means more clouds, and clouds can act both as a blanket over the land or as a reflector of solar energy. Which will be its dominant influence in the future is one of the global warming mysteries.

The Least You Need to Know

♦ The greenhouse effect is real and we need it, otherwise we'd freeze.

♦ Carbon dioxide, methane, nitrous oxide, chlorofluorocarbons, and some other long-lived gasses may be adding to the effect.

♦ Water vapor acts as a greenhouse gas, but it's short lived.

♦ Man's activities are upsetting the delicate balance of these elements.

The Universe Was a Gas

In This Chapter

- ◆ The big bang got it started
- ◆ The evolution of gas in the stars
- ◆ How the atmospheres of the planets differ
- ◆ How our current atmosphere evolved

To understand the delicate balance of the atmosphere, it's a good idea to understand how it all came to be. In this chapter, I'll discuss the evolution of atmospheric gas in the universe and on Earth from the big bang to global warming.

First Thoughts

Not only has our atmosphere evolved, but our perception of it has, too. When ancient astronomers looked out into space they thought they were looking at the inside of a big dome—like the Astrodome, only bigger. Their idea was that Earth consisted of Europe, Asia, and Africa surrounded by a huge circular body of water called the River Ocean. Anaximenes, an early Greek philosopher (d. 528 B.C.E.), projected that everything was made of air, only in different densities. When it was thin, it was fire. When it condensed it became wind and cloud; more condensed, it became water, earth, and stone. Aristotle (384–322 B.C.E.), like most other philosophers of his time, felt the world was made of air, fire, earth, and water. His model of the universe showed the planets and the stars rotating around the earth.

It was a long time before Copernicus in the 1500s put the idea of the earth as the center of the universe to rest when he described the planets revolving around the sun—not around the earth. Copernicus argued that the motion we see in the heavens is the result of the earth's daily rotation on its axis and yearly revolution around the sun.

That new theory, however, required a drastic change in man's perception of the universe beyond the planets. For one thing, it had to be a lot bigger than astronomers had previously conjectured. Otherwise, why did the stars always appear in the same positions and at the same brightness?

Bang

It took until the beginning of the twentieth century for astronomers to really get a fix on how big the universe was and what it was made of—mostly gas. U.S. astronomer Edwin Hubble was the first to discover other galaxies. Not only were we not at the center of our galaxy, but our galaxy was only one of billions of galaxies. He also figured out that they were moving away from us … and each other. In other words, the universe was expanding. Hence, scientists concluded that the universe must have very long ago been compact and dense. And the explosion of that primary matter was the birth of the universe, an event that today we call the *big bang*.

Warm Words

The **big bang** is a theory that the universe we know began its existence about 13 billion years ago as the explosion of a small super-dense concentration of matter, transforming it into gas.

I visited the Yerkes Observatory in Williams Bay, Wisconsin, not too long ago, and got to view some of these stars with astronomer Kyle Cudworth, director of the observatory. It was in December, and a series of winter storms rippled through the atmosphere above Lake Geneva. Cudworth took advantage of a break in the storms to look through the observatory's great 40-inch Yerkes telescope at the observatory built by Alvan Clark and Sons. At the beginning of the twentieth century, this was the largest telescope in the world.

The dome where the telescope is housed looks like a cathedral. In the middle of the cavernous room, balanced on a giant metal pier, is the telescope. It is 63 feet long with a magnifying power, depending upon the eyepiece, of between 200 and 1,000 times that of the unaided eyes. The telescope looks regal, the aged but elegant former king of astronomy. Both Albert Einstein and Edwin Hubble once put their eye up to this telescope through which I then viewed. What I saw was M15, a globular star cluster that orbits just outside our galaxy. Do any of these stars have planets? And how many of those planets have problems with the global warming?

Astronomers looking through Clark telescopes like this discovered Pluto, the fifth moon of Jupiter, the spiral arms of our own galaxy, and the fact that other galaxies were moving away from us—the first evidence of an expanding universe. Telescopes have taught us a lot about our universe and the evolution of materials that make up our atmosphere.

First Gas

NASA scientists looking through today's space-based Hubble Telescope at some of the oldest objects in the universe believe the universe is about 13 billion years old. That's when many believe the big bang occurred. Initially the universe was pure energy, but some of that immediately changed into matter. Most of the original material was hydrogen gas. As it expanded, it began to clump together. Most of the material in the universe remains hydrogen gas, though it is so light that little of it remains in Earth's own atmosphere.

Stars Form

Stars begin their lives as swirling clouds of gas and dust. As these swirls come to-gether they reach a point of *critical density* where the gravity of the individual particles becomes strong enough to draw them together. As the cloud comes together it shrinks up and gets denser and hotter at the core until the region is super hot. At this point nuclear fusion occurs as hydrogen atoms collide with one another, fusing into helium atoms.

Warm Words

Critical density is that point in the evolution of a star where gas and dust come to-gether close enough for gravity to take over and bind the elements together.

Newborn stars gradually settle down into their adult phase, cranking out energy in the form of heat and light from this conversion of hydrogen to helium. These two gaseous elements are the dominant elements in most first generation stars. After a star has used up its hydrogen, it will begin to convert helium to heavier elements like oxygen and carbon. These building blocks will be used much later to create carbon dioxide and other greenhouse gases.

When a star has exhausted all of its fuel, it expands and may become a red giant or a super giant. Eventually some of these massive stars become supernovas and explode. In these explosions the heavier elements are made, including uranium. These elements are thrown back into the universe and again become part of the gas and dust, which may lead to second-generation stars. Astronomers see very few first generation stars any more. Most are second to tenth generation. The exploding supernovas create the stardust or elements of which we are made.

Climatoids

Most stars get their energy through the process of nuclear fusion. Four hydrogen atoms come together to create a single helium atom, a reaction so intense it appears as fire. Our own sun converts about five million tons of matter each second into energy. So far it's used about half of its available hydrogen fuel. At that rate it will take another 4.6 billion years to run out of gas.

Climatoids

In the song "Woodstock" by the late '60s rock band Buffalo Springfield, which then included Neil Young, was a lyric that claimed "we are stardust." Whether the words were a mystical notion or scientific inspiration is uncertain, but the idea is correct. All the elements in our bodies and in the atmosphere were created within the fiery furnaces of the stars.

Our Solar System

Our solar system was born about 4.6 billion years ago, a cloud of stardust that contracted, perhaps the effect of a nearby supernova explosion. The supernova explosion enriched the cloud with various elements including silicon, iron, and carbon. As those elements came together they began to rotate faster, getting hotter in the process. Most of this mass gravitated toward the center of that swirling cloud and congealed into our sun. The gas and dust outside the sun formed the planets, which orbited around it.

There was still plenty of debris in this solar orbit, but over time much of it collided with the new planetary bodies and became part of their masses. Planets like Mercury or satellites like our moon, which lack an atmosphere to erode those impact points, bear colossal craters on their surfaces to attest to these collisions.

About a billion years of this kind of treatment resulted in the major planets of our solar system. In order of appearance from the sun, they include Mercury, Venus, Earth, Mars, Jupiter, Saturn, Uranus, Neptune, and Pluto. Seven of the major planets have satellites; some of the satellites are themselves planet size. Plus there are lots of asteroids, comets, and meteors out there—the remnants of the original stardust that got us going. None of them have atmospheres.

There are two types of planets in our solar system, *terrestrial* and *gas giants*. The gas giants are massive planets with low densities and thick atmospheres formed mostly from hydrogen and helium. Neptune, Uranus, Saturn, and Jupiter are what we refer to as gas giants. They are thought to have rocky cores, although no one knows for sure. Terrestrial planets are smaller and denser. They are made of rock and have thin atmospheres or no atmosphere at all. Mars, Earth, Venus, Mercury, and Pluto are terrestrial planets.

Warm Words

Terrestrial planets are the smaller rocky planets Earth, Mars, Mercury, and Venus, which might or might not have thin atmospheres. **Gas giants** are the huge gaseous planets Jupiter, Saturn, Uranus, and Neptune, which have incredibly thick atmospheres.

Here Comes the Sun

The sun is certainly the most conspicuous body in our solar system. Without its energy there would be no life on Earth. Most ancient religions have a powerful sun god, and the winter solstice was the scariest celestial event in many primitive cultures. That's because when the sun sank to its lowest point in the southern sky, sky watchers and shamans feared it wouldn't return. So many of these early Native American societies built astronomical observatories where ceremonies were enacted on the winter solstice to ensure the sun stuck around.

So far as stars go, the sun has only average size, mass, and brightness. It takes sunlight about eight minutes to reach our planet, a distance of about 150 million kilometers. The next nearest star is Proxima Centauri, and its light takes about 4 years to reach us. The sun is about 4.6 billion years old. That's somewhere around mid-life crisis, so far as stars go. Most of the mass of the sun is a huge gaseous wrap around its core, running at a temperature of about 5,500K, turning the sun's surface yellow-white. It's not the kind of atmosphere you'd want to raise your kids in.

The surface of the sun is a patchwork of turbulent cells, each about the size of Texas, that brings up energy from the core to the surface. The surface of the sun is anything but placid, with huge loops of gas called *prominences* towering above it. There are also violent eruptions that occur which are called *solar flares*. These events spew light and charged particles into the air. Those particles can cause *auroras*, bad TV reception, and even short circuits in your local power grid.

Strong magnetic fields on the sun's surface can create dark areas know as *sunspots*. Sunspot activity is cyclical with peaks occurring every 11 years. Between the years 1645 and 1715 C.E., during the peak of the Little Ice Age, known for its consistent cold weather, there were almost no sunspots. Currently we are in a period of high sunspot activity. Some scientists wonder at the relationship with respect to our present global warming.

Mars

Mars is known as the Red Planet because of its rust-red hue caused by iron oxide or plain old rust. It is our closest planetary neighbor, and has been the focus of many of our unmanned planetary explorations in recent years. This is in part because Mars is the closest in appearance and structural makeup to Earth, and scientists often wondered if life, even in its simplest forms, ever evolved on Mars.

In the late nineteenth century, Italian astronomer Giovanni Virginio Schiaparelli described what he called *canali* on Mars, what some people speculated might have been canals built by a dying Martian civilization to ship water around the planet. Percival Lowell, a wealthy industrialist, mathematician, and amateur astronomer, built an observatory in Flagstaff, Arizona, in 1894 and spent numerous nights observing and describing those canals. However, photographs taken from exploratory spacecraft sent to Mars revealed no canals or anything that even looks like canals. Many feel that what Lowell really proved was that if you stay up long enough staring through a telescope you might end up seeing things.

The idea that life may have one time existed on Mars is not complete science fiction. The atmosphere of Mars is currently only about 0.7 percent as thick as the earth's, but scientists think they might have observed various remnants of an ancient atmosphere in the space surrounding the planet. It's possible that primitive forms of life existed on the planet about 3 or 4 billion years ago when Mars may have had an atmosphere similar to ours. But Mars is smaller, has less gravity to hold onto its atmosphere, and less volcanic activity to recycle atmospheric gases locked in the soil.

Climatoids

The idea that there might be Martians running around on the Red Planet was a part of the popular culture in the early twentieth century. So much so that widespread panic greeted Orson Welles's 1938 radio broadcast of the "War of the Worlds," a fictional account of a Martian attack.

Climatoids

The size of a planet has a lot to do with whether or not it can generate enough gravity to hold on to its atmosphere. Mars was too small to hold on to its gases. But Jupiter is far more massive and has held on to a lot more gas than Mars or Earth could contain.

Mars has seasons, just like Earth. Mars has polar icecaps, but they are a combination of water ice and dry ice, or frozen carbon dioxide. The changing of the seasons increases the prevalence of dust storms as well as the size of the polar caps. The climate ranges from a high of 60°F (15.5°C)—at noon on a hot day at the equator—to a low of 225°F (204°C below zero). It's the kind of thing you end up with when you *don't* have a greenhouse effect.

Jupiter

Jupiter is the largest of the planets in our solar system. It's also a whirling dervish, spinning once on its axis every 10 hours. Jupiter has 16 satellites. The four most prominent are Europa, Io, Ganymede, and Callistro. Scientists are particularly interested in both Io and Europa. Europa has a smooth surface that is probably water ice. It could be the most hospitable to life in the Jupiter system. Io is also interesting because it has some of the most magnificent volcanoes in our entire solar system. Magma, mainly molten sulfur, is thrown hundreds of miles into space, far above its thin atmosphere. The low gravity of Io may enable the highflying fireworks displays.

Jupiter itself has an atmosphere mostly of hydrogen and helium, the two basic elements of the universe. The brightly colored surface on Jupiter comes from trace elements of methane and ammonia. Storms and atmospheric disturbances are common on Jupiter's surface. One such storm, The Great Red Spot, stretches over the surface of Jupiter in an area larger than the planet Earth and has been blowing for more than 300 years.

Venus

If you want to get the feel of what a run-away greenhouse effect might give you, you have only to look as far as Venus. Venus is the third-brightest object in the sky. It's closer to the size and gravity of Earth, so it retains its atmosphere. The Greeks associated the planet with Aphrodite, the goddess of love. Venus is the goddess's Roman name. The symbol for Venus is commonly used as the symbol for women or the female sex. The International Astronomical Union is considering naming the craters on Venus after famous women. Gertrude Stein, Pearl Buck, and Margaret Mead are current candidates.

The atmosphere on Venus is very thick compared to Earth, though not as thick as Jupiter. Venus has an atmospheric pressure that is 90 times what you experience on Earth's surface. You can't see the surface of Venus as it is covered with clouds. The clouds are mostly sulfuric acid, the atmosphere mostly carbon dioxide (about 96 percent) with a little nitrogen. What does that get you? A runaway greenhouse effect. At the top of those clouds it's below freezing, but on the surface of Venus it's more than 900°F (500°C). Now that's what we call global warming.

The greenhouse effect causes only a mild 60°F (33°C) rise in the surface temperature of the earth, but the greenhouse effect raises Venus's surface temperature by almost 900°F (500°C). Though Venus is closer to the sun, that's not the reason it's so hot. The clouds reflect 75 percent of the sun's rays. However, the greenhouse effect on Venus is so efficient that it prevents all but a negligible portion of the heat generated at the surface from escaping into space.

Earth

Earth is the prototypical "terrestrial planet," small, rocky, with a thin atmosphere, as opposed to the voluminous atmospheres of the gas giants. Earth is a layered planet. It has a solid core at the center, a molten core surrounding that, a mantle, and a crust upon which the land masses and the oceans rest. Scientists believe that the crust of the earth is divided up into various plates, which drift in relationship to each other. The theory is called *plate tectonics*.

> **Climatoids**
>
> **Plate tectonics** is the theory that the earth is made up of plates that move around, bump, and collide with each other in ways that create some of the earth's most spectacular land forms. The Himalayan Mountains were created when the floating land mass that was India plowed into southern Asia. That's how Mount Everest and its other giant neighbors were born.

Earth's atmosphere consists mostly of oxygen and nitrogen. But this stable mixture didn't always exist. Our current atmosphere is the result of a very gradual evolution that began soon after the earth was born, about 4.6 billion years ago.

Volcanic Earth Spews Carbon Dioxide

The first atmosphere, the one that was around shortly after the planet formed, was probably swept away by solar winds, vast streams of particles emitted by the sun. But as the earth cooled, a solid crust was formed and the gases dissolved in the molten rock were gradually released. The atmosphere was then very similar to what is spewed into the air by volcanoes. The principle components of this new atmosphere were probably water vapor, carbon dioxide, and nitrogen.

As the earth continued to cool, the clouds formed and the great rains began to fall. At first the rain evaporated before hitting the surface of the earth, but the process accelerated the cooling, and eventually the rain reached Earth and began to fill the oceans, rivers, and lakes. This took a lot of the water vapor from the air, and a fair amount of carbon dioxide as well. However, carbon dioxide was still the dominant component in the air.

But how did the oxygen-rich atmosphere that we now have come into existence? The volcanic processes that existed on the early planet don't create oxygen. Some water vapor (remember, H_2O) is carried into the upper atmosphere where the sun's ultraviolet radiation separates it into hydrogen and molecular oxygen (O_2) but this is a very slow process and can't account for all the oxygen we see in the atmosphere today.

Life Changes It to Oxygen

The main source of oxygen, it turns out, is life itself in the form of green plants. By the process of photosynthesis, plants use sunlight to change water and atmospheric carbon dioxide into organic matter, and in the process release oxygen back into the atmosphere. Scientists believe that the first form of life on Earth may have been bacteria that carried out their metabolism in the absence of oxygen. Today some of those *anaerobic* bacteria still exist, though most animal life—birds, insects, reptiles, mammals—are *aerobic*, requiring oxygen to metabolize.

Scientists estimate that the earliest photosynthesizing microbes bloomed in the sea some 2.8 to 3.5 billion years ago. At that time the atmosphere contained perhaps 1,000 times more carbon dioxide than it does today. But it was the evolution of plants that supplied most of the oxygen that eventually led to our current atmosphere, which supports the higher forms of life. With plants giving off oxygen, slowly the percentage of oxygen in the atmosphere increased. By the beginning of the Paleozoic Era, some 570 million years ago, the fossil record indicates that oxygen-requiring organisms were abundant in the sea. From a primordial envelope of nitrogen and carbon dioxide, it had changed to a mixture of about 78 percent nitrogen, 21 percent oxygen, and a trace of carbon dioxide and other gases. Two million years later, during the Devonian Period, land plants became widespread.

> **Warm Words**
>
> **Anaerobic** means living or occurring without oxygen. **Aerobic** means living or occurring in the presence of oxygen.

Man Changes It Back?

The earth's atmosphere has evolved over time from one that was free of oxygen to one that had significant levels of free oxygen. In the process, a large portion of the carbon dioxide that was originally mixed into the atmosphere during volcanic eruptions has disappeared. The question is, are we now returning the atmosphere to one where carbon dioxide will again play a dominant role, as it did before the evolution of higher animals? Right now, man is engaged in a colossal experiment—dumping tens of billions of tons of carbon dioxide and other heat-trapping gasses into the air. Humans have a history of adapting to shifts in the climate. Only the shift we are now facing might come more rapidly than any we've experienced before.

The Least You Need to Know

- The original gases that made up most of the primitive universe were hydrogen and helium.

- Nuclear reactions within the stars and tremendous explosions at the end of a star's life convert simple gases to more complex elements.

- We all are made of stardust.

- The solar system is comprised of rocky planets and gaseous ones.

- The atmosphere on earth is the evolutionary product of volcanic gas (CO_2) converted to oxygen, primarily through the photosynthesis of plants.

Chapter 4

Is That a Hole Up There?

In This Chapter

- The atmosphere is like a layer cake
- How the ozone layer protects us
- How we almost destroyed the ozone layer
- The world bans CFCs

Do you remember in the preceding chapter when I was explaining how plants in the sea first started creating oxygen? Well, a portion of that oxygen rose to some of the highest regions of the planet's atmosphere where ultraviolet radiation (UV) from the sun caused a reaction that led to the forming of O_3 or ozone. That ozone settled into a thin layer that effectively blocked out deadly doses of solar UV radiation, which until then had bombarded the surface of the earth. It was only with this atmospheric shield that animal life, which first developed under the protection of water, was able to crawl out on land. In this chapter, I'll tell you all about it.

The Layers of the Atmosphere

To understand the science and the controversy surrounding the ozone layer, you have to understand a little bit about the multiple layers of our atmosphere and how sunlight reacts as it passes through them.

The *thermosphere* is the top layer of the atmosphere. It extends up to 400 miles or more above the surface of the earth. As the sun's rays enter this level, the ultraviolet (UV) portion of the spectrum interacts with the atmosphere, which at that height is pretty thin. But what's up there quickly grabs the heat in those rays, which is why the thermosphere is quite hot at higher altitudes. It cools, however, as it descends toward the *mesosphere*.

Climatoids

Though the temperature in the outer limits of the thermosphere can heat from UV radiation, the air there is so thin that the fast-moving air molecules, though hot, seldom collide with anything. If an astronaut were to stick his hand out into the hot part of the thermosphere, his hand would not feel "hot."

As the sun's rays hit the mesosphere, the next layer of our atmospheric cake, the temperature reverses and starts rising again. Even though the air is still thin here, there is enough to absorb the rays and restart the heating process.

Below the mesosphere is the stratosphere. In the stratosphere we start getting cold again. But the drop isn't as big as it is through the thermosphere. The weather through the *stratosphere* is closer to that of the earth's surface.

Warm Words

The **troposphere** is the bottom layer of the atmosphere, rising from sea level up to an average of about 7.5 miles (12 kilometers). The **stratosphere** is above that and rises to an average of 31 miles (50 kilometers). The **mesosphere** lies on top of that and runs upward to about 50 miles (80 kilometers). The **thermosphere** tops the atmosphere off, rising to around 400 miles (644 kilometers), though it really has no well-defined upper limit.

Climatoids

In the formation of a hurricane, the eye of the hurricane acts as a chimney from which heat is exhausted into the stratosphere. Only in tall violent storms does the troposphere react with the stratosphere.

As the sun's UV rays finish their descent through the stratosphere, they encounter the *troposphere*. That's where we live. It's the bottom of the atmosphere and it extends from sea level to about 7.5 miles (12 kilometers) high. When the sun's UV rays hit the troposphere, the air is pretty cold, less than -60°F (-51°C). But the UV rays of the sun hit the surface of the earth and radiate back infrared radiation (heat). This is where the greenhouse gases do their good work by absorbing that heat and making things comfortable. The troposphere is also the layer where all our weather takes place. It's generally where the clouds stop. Seldom can they penetrate the stable air of the stratosphere, except in violent storms.

The Ozone's Vital Role

But what happened to that ozone layer, you ask? Well, it's back up in the stratosphere, the layer above the troposphere where we all live. The stratosphere contains about 97 percent of all the ozone in the atmosphere. Oxygen (O_2) rises from the earth's surface up into the stratosphere, where it reacts with the sun's incoming UV rays, goes through a series of chemical changes, and comes out as ozone (O_3). This layer of ozone encounters the UV rays, breaks up, and reforms in a continual process that absorbs the UV rays. It's like a celestial pool game with one ball connecting to another and then another, each connection taking up more UV rays.

This chemical pool game protects us from the extreme UV rays that enter the stratosphere. We still get some UV rays, but not the dose we would get without the ozone layer. In fact, if those UV rays passed through the stratosphere unimpeded, they'd take us out. One of the reasons scientists feel that Mars is lifeless today is because it's lost its UV shield, and that makes the place currently uninhabitable.

Billions of years ago there was no ozone layer to protect the earth from the sun's UV radiation. Scientists believe that life only existed in the sea, protected by the water. However, a layer of ozone gradually developed in the stratosphere. It was only about 600 million years ago that this layer grew thick enough to trap most of the UV radiation, and the land could then support life.

> **Climatoids**
>
> The greatest absorption of the UV rays takes place between 25 and 50 kilometers above the earth. The peak concentration of ozone (03) occurs at around 25 kilometers, the infamous ozone layer.

> **Global Warnings**
>
> To get an idea what could happen on Earth without the ozone layer, look to Mars. UV radiation there bombards the planet's surface, creating reactive superoxides and peroxides that kill any available organic molecules. As Carl Sagan put it, "The surface of Mars is antiseptic with a vengeance."

How the good ozone is made.

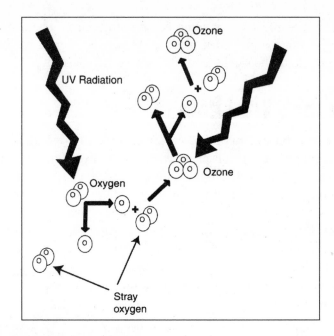

A Strange-Smelling Chemical

Ozone was first discovered in 1840 by a German chemist, Friedrich Schönbein. It's said that he was working in a poorly ventilated laboratory and became conscious of a peculiar odor in the neighborhood near electrical equipment. He tracked it down and discovered a gas, which he called *ozone*, which is from the Greek word for "smell." Thomas Andrews, an Irish chemist, later identified it as a form of oxygen.

Its presence in the upper atmosphere was deduced at first theoretically by the absence of certain wavelengths of solar ultraviolet radiation at ground level. Scientists deduced the absence could have only occurred if the radiation were intercepted by ozone. Later, rockets and satellites confirmed this initial finding.

Warm Words

Ozone is a gaseous, almost colorless form of oxygen that protects the earth against ultraviolet radiation in the upper atmosphere, but is part of the chemical pollution nearer to Earth.

The first estimates of the abundance of ozone in the atmosphere were made in 1913. Systematic observations all around the world began in 1930. But it wasn't until 1974 that the world's attention was focused on the ozone layer. Then chemists F. Sherwood Rowland and Mario Molina of the University of California at Irvine announced to the world that the ozone layer was under attack by a family of commonly used industrial compounds known as chlorofluorocarbons (CFCs).

The Ozone Depletion Story

To understand the ozone depletion story, you have to understand something about CFCs. CFCs were a family of industrial compounds created in 1928 by an eccentric Ohio scientist named Thomas Midgley Jr. CFCs are the chemicals that make your refrigerator work, and in the days when everybody was still waiting for the iceman to cometh, that discovery was a big deal. CFCs were thus the miracle chemical of the day. Plus CFCs had a lot of other uses as well. They were the forces behind aerosol sprays. They were used as fillers in insulation. They were also used in air conditioning units to cool the home and the car.

The reason that CFCs did their job so nicely was that they were stable chemicals. Nothing scientists had tried yet could break them up into their constituent atoms of chlorine, fluorine, and carbon. Before the 1970s these chemicals were thought to be harmless. But the guys that said they were harmless were wrong.

In the early 1970s, Mario Molina, a chemist who'd worked on lasers at the University of California at Berkeley, joined a group under F. Sherwood Rowland at the University of California at Irvine. Molina's new job was to track down CFCs which had been accumulating in the atmosphere for some time. His first task was a systematic search for processes in the lower atmosphere that might destroy them, but CFCs were so tough, nothing seemed to affect them. So the question was, where had they gone?

Hot Debates

Mario Molina was born in Mexico City. He attended the University of California at Berkeley in the 1960s, where he got his doctorate and worked with chemical lasers. He was dismayed by the fact that they could be used for military purposes as he'd been a peace activist at Berkeley. So when he moved to Irvine, California, and joined the group of F. Sherwood Rowland, Molina chose as his research project the benign goal of finding out the environmental fate of certain inert industrial compounds called CFCs which had been accumulating in the atmosphere for some time. Molina had no idea of the controversy he was about to get involved with.

Molina and Rowland determined that the chemicals would eventually drift into the upper atmosphere where they could be destroyed by solar radiation. The problem was the chlorine atoms produced by the decomposition of CFCs could destroy ozone. The continued release of CFCs into the atmosphere could cause a significant depletion of the ozone layer, the chemicals in the stratosphere that protected Earth from excessive UV rays.

The two published their findings in the June 28, 1974 issue of *Nature*. But they didn't stop there. Molina and Rowland knew that the information could die if its only exposure was to the scientific community through their paper. So in the years following publication, the two scientists took to the road and the air to spread the message to policy makers and the news media. They continued to publish articles on the CFC/ozone issue and presented their results at scientific conferences and testified before legislative hearings where lawmakers were considering controls on CFC emissions.

They projected a decrease in the ozone content in the atmosphere that could have serious health effects. These gasses had been around for 50 years, so a lot of damage had already been done, and if we didn't stop putting them into the atmosphere, it was going to get worse. And even if we ceased manufacturing these gases completely, it would take another 100 years for the atmosphere to return to preindustrial conditions.

Molina and Rowland set into motion a contentious debate between scientists, policy makers, environmentalists, and industry that was to last for 13 years. CFCs in aerosol sprays were banned in the United States in 1978, but many people remained unconvinced of their danger. Why eliminate the use of a vital chemical and a multimillion-dollar industry on hypothetical evidence?

The Price of Too Much UV

So what is all the fuss about? Molina and Rowland weren't talking about an end to the ozone layer; they were talking about a reduction of 3 to 7 percent. What's that? A little more UV radiation? Don't people go to tanning salons for that sort of stuff? But even this level of ozone depletion would bring serious health consequences.

Effects on Humans

According to the United Nations Environment Program, increased UV radiation can impact our immune system and lead to the spread of infectious disease; increase the risk of skin cancer, particularly basal and squamous cell carcinomas and the more deadly melanoma; as well as cause cataracts. More specifically, increased UV radiation affects our immune system, skin, and vision as follows:

◆ **Immune responses** The evidence continues to mount that UV radiation can suppress immune reactions in both laboratory animals and humans of all skin types. UV radiation can impair certain antibodies. Studies show that UV radiation can not only decrease the immune responses to infectious organisms, but can increase the severity and duration of a number of infectious diseases. And it can not only increase the severity of infectious diseases, but it can also decrease the effectiveness of vaccinations.

♦ **Skin cancer**   UV rays are known to cause skin cancers. However, a recent study estimated that in the absence of the Montreal Protocol, elevations in UV due to ozone depletion would cause a four-fold increase in skin cancer by the end of the twenty-first century. Even under current restrictions, the relative increase in skin cancer is still estimated to increase by 10 percent until 2060. After that it should gradually return to preindustrial levels. That requires that everyone follow the rules, though.

♦ **Vision**   Recent experiments show that rats who are exposed to UV rays develop the precursors to cataracts at the periphery of the lens.

Furthermore, UV radiation can seriously impact our Earth and its aquatic ecosystems.

Effects on the Earth

UV radiation can have serious effects on plant flowering, pollen development, seed production, and seed size. UV rays can also promote plant pests, including various microorganisms and insects. It can even damage the plant's DNA. Plants exposed to excessive doses of radiation show a variety of deformities like shorter stems, multiple branching, and other effects.

There are some positive effects. Some Mediterranean species developed thicker wax layers on leaf surfaces under elevated levels of UV radiation; this helps the plants retain more water during the dry summer months.

Effects on Aquatic Ecosystems

Oregon State University researchers discovered strong evidence in the 1990s that the decreasing number of amphibians, especially toads and frogs, throughout the world is related to an increase in UV radiation. The researchers hypothesized that the increased UV destroys amphibian eggs. Serious declines in the world's coral reefs may also be an effect of increased UV rays.

Phytoplankton are tiny plants found in the top sunlit portion of the ocean which drift with the currents. They are the basis of most of the ocean's food chain. Krill eat phytoplankton and a number of animals eat krill, including penguins, seals, and humpback whales. Though UV radiation might not harm humpback whales directly, it will kill phytoplankton and that will destroy an important part of the food chain upon which the whales depend.

Warm Words

Phytoplankton are microscopic plant life found drifting in the upper sunlit portion of the sea or freshwater, and are the basis of the food chain for other marine organisms.

Phytoplankton are also a major consumer of atmospheric carbon dioxide. Thus, the destruction of phytoplankton could have immediate effects on wildlife, as well as lead to increases in global warming.

The Antarctic Ozone Hole

In the early 1980s, a lot of this was still speculative. But then in 1985, the British team of Joseph Farman and colleagues discovered a large hole in the ozone layer over Halley Bay Station in Antarctica. When the team first measured ozone above the station, they assumed that the spectrophotometer they were using must be broken and sent the device back to England to be repaired. But the spectrophotometer worked just fine.

In the summer of 1986, NASA sent National Oceanic and Atmospheric Administration senior scientist Susan Solomon and a team of 16 scientists to Antarctica to investigate. Solomon was only 30 years old at the time but had already co-authored a standard textbook in her field and had received an award from the American Geophysical Union.

Climatoids

When CFCs are broken down in the atmosphere, they liberate chlorine atoms. One chlorine atom is capable of destroying 100,000 ozone atoms.

Solomon's breakthrough was the discovery of a critical link between manmade CFCs and the ozone hole above Antarctica. It was Solomon's theory that the ice crystals in the clouds over Antarctica might assist the breakdown of CFCs into reactive chlorine. When the depletion was verified, it came to be known throughout the world by a series of NASA satellite photos as the Antarctic Ozone Hole.

For the public, the Antarctic Ozone Hole was the smoking gun. It was the first time anyone had ever presented solid, scientific, irrefutable evidence of the connection between CFCs and ozone depletion. The public began to realize man's enormous capacity to affect his environment. In the case of the Antarctic Ozone Hole, we had created a huge, new, semi-permanent feature on the planet that was bigger than the United States.

Scientists got busy checking elsewhere. A series of careful measurements taken from old converted U-2 spy planes as well as the Nimbus-7 satellite indicated that ozone depletion over North America and Europe was taking place at a much faster rate than was previously anticipated. Declines were as rapid as 0.5 percent a year. That figure meant that the time when ozone depletion could have serious effects to health and the environment was shortened from a century to decades.

Global Warnings _____

By the mid-1980s, scientists were also uncovering problems with greenhouse gases. They discovered that CFCs were not only dangerous to the ozone layer, but were an efficient greenhouse gas as well. Just like carbon dioxide, methane, and nitrous oxide, CFCs contributed to global warming by absorbing some of the infrared heat reflected back from the surface of the earth. With the bad press mounting, it seemed that CFCs were doomed.

The World Fights Back

In 1987, the industrial nations gathered in Montreal, Canada, to discuss the crisis. Though there were still members of the chemical industry that urged restraint, few of the international representatives were willing to take the risk in light of the mounting evidence. In the end, 25 nations signed the agreement to set limits on the production of CFCs and other related substances. By that initial agreement, there was to be a gradual phase-out of CFCs and other ozone-depleting chemicals. It was the first global environmental accord. Today, over 168 nations are a party to that agreement.

Climatoids _____

In April 5, 1988, President Ronald Reagan issued a statement praising the agreement: "The Montreal protocol is a model of cooperation. It is a product of the recognition and international consensus that ozone depletion is a global problem, both in terms of its causes and its effects. The protocol is the result of an extraordinary process of scientific study, negotiations among representatives of the business and environmental communities, and international diplomacy. It is a monumental achievement."

The initial agreement called for a gradual phase-down of production and consumption of CFCs and related chemicals, but as the bad news continued to mount from the scientific arena, a sense of urgency grew. At subsequent meetings in London, Copenhagen, Vienna, and again in Montreal (all in the 1990s), the phase-out was accelerated. Most CFCs were banned by 1996.

Today, scientists have developed a number of alternatives to CFCs, and more are being developed. Scientists report that the effort appears to be worth it. The ozone layer is already showing signs of recovering. Special instruments aboard satellites have

The controversy over the validity of claims by chemists Mario Molina and F. Sherwood Rowland of the University of California at Irvine that CFCs were depleting the ozone layer came to an end in 1995 when they won the Nobel Prize for their discovery. It was the first Nobel Prize ever given for an environmental science.

shown that the amount of chlorine in the stratosphere peaked in 1997 and appears to be on the decline. The ozone hole over Antarctica grew to its maximum size in 1998, and was a little smaller in 1999. It appears to be oscillating around this high value but is expected to go down. Scientists predict that it may not be until the middle of the twenty-first century before the effects disappear.

Rowland and Molina, who were initially hounded for their predictions in the mid-70s, were vindicated when they received the Nobel Prize for their discoveries.

Scientists Look Elsewhere

Perhaps the biggest prize of all was the global cooperation that this issue seemed to engender. Despite extreme opposition from the chemical industry, lots of controversy in the scientific community, and the challenge of fashioning an agreement that would satisfy the needs of both the industrialized and nonindustrialized world, the world came together and negotiated an agreement that for the first time tackled an important environmental issue.

Under the excitement of this important moment, many now looked to the problem of global warming and hoped for similar cooperation and a similar result. But the going hasn't been as smooth. Banning CFCs involved controlling the production of a single family of industrial chemicals, but the causes of global warming are much more numerous.

In the end, it's gotten down to the fact that only a few dozen companies worldwide produced CFCs. According to one negotiator, you could put them all in one room and talk to them. But you can't do that for the producers of greenhouse gases. Imagine the size of a room that would hold representatives from all the world's utilities and industries. Carbon dioxide, methane, and the other greenhouse gases are the byproducts of the very heart of modern civilization—including industry, transportation, power, and agriculture. It's a much bigger ball game.

Ozone Down Here

Ozone in its natural form is not only present in the stratosphere but also at ground level, mostly from ozone manufactured in the stratosphere that has floated down. But in the twentieth century, another type of ozone has emerged. In the presence of hydrocarbons and nitrogen oxides coughed from the exhaust of automobiles, ozone is produced through a series of complex chemical reactions, which lead to *smog*.

The origins of the word smog may have come originally from a combination of fog and smoke in old England, but more recently it's come to mean the polluted haze found in many American cities, particularly in the southwestern part of the United States. It's not really smoke and fog, but a chemical stew of pollutants (primarily from automobiles) that are transformed to smog by a reaction with the sun.

Warm Words

Smog is pollution and comes in two types. The gray smog of older industrial cities like London and New York is derived from the massive combustion of coal and fuel oil in or near the city. The brown smog characteristic of Los Angeles and Denver in the late twentieth century comes from automobiles.

Traffic, the twenty-first century ozone machine.

(Photo by Michael Tennesen)

Smog's primary ingredients are nitrogen oxides and volatile organic compounds, which when mixed together create ozone. That's right, the ingredient that's so vital to humankind up in the stratosphere is real nasty when encountered down here on Earth. There are a number of other chemicals within smog that are dangerous, but ozone seems to get all the bad press.

Climatoids

Elderly people in Los Angeles, Denver, and other southwest metropolitan cities affected by smog frequently go to malls early in the morning to walk or exercise. The air-conditioned climate protects them from the damage of ozone and is supported and encouraged by businesses and communities alike.

Ozone can irritate the eyes, nose, and throat. Ground-level ozone poses a significant health risk to people who suffer from chronic lung disorders such as asthma and bronchitis. Lungs will age a lot faster in an environment with ozone, even at low levels. It's also bad if you exercise, work, or recreate outdoors.

Ozone is not only bad for your health but it can ruin your tires, take the shine off your paint job, make your nylons run, and wreck your clothing. Plus, it's nasty to crops and plants. Some people estimate the damage to crops in the United States at about $1 billion a year. It also attacks the cool moist leaves in trees and can contribute to forest fires.

The Least You Need to Know

◆ Our atmosphere is a multi-layered affair.

◆ Ozone in the stratosphere protects life on Earth from a nasty dose of ultraviolet radiation.

◆ Scientists in the 1970s discovered that the ozone layer was under attack by CFCs, which were commonly used for air conditioning and refrigeration.

◆ The world got together in Montreal and agreed to phase out CFCs and their sister chemicals.

◆ Ozone may be beneficial in the stratosphere, but down here on Earth in the form of photochemical smog it's just plain unhealthy.

Part 2

A Historical Perspective

So now you've discovered that greenhouse gases are increasing in the atmosphere and it looks like our machines are putting them there. What do we do next? To answer that question, scientists have had to look into the past to see how these gases have varied over geological time and what the consequences were for the atmosphere.

To understand that past, I'll invite you to the Greenland ice cap and look at samples of air locked in the ice that go back 110,000 years. Those cores show us that this modern era of climate stability may be a false prophet. The atmosphere is really a roller coaster and wide swings in temperature are the norm rather than the exception.

I'll take a look at how the climate affected the Mayans, the Vikings, and what the Little Ice Age that peaked in the early 1700s did to us all. Then I'll talk about how industrialization and deforestation are affecting our atmosphere right now.

Taking a Reading

In This Chapter

- How we record the world's temperatures

- Extracting ancient air from Greenland ice

- What else the ice tells us

- Sea bottoms, tree rings, and other ways to read the past

To figure out how to tell the future you need to be able to read the past. The past tells you how temperatures have changed, what the causes were, and what the effects were. The trouble is we haven't been taking the temperature that long. Scientists have had to come up with a few sly methods to tell us what the weather was like in the past.

The Historical Record

Moms weren't able to read their kids' temperatures before Galileo. The same guy that popularized the telescope invented the thermometer in the early 1600s. The barometer and the rain gauge followed a little later. The big problem was standardization. One scientist, comparing and calibrating various types of thermometers in the late 1700s, listed no fewer than seventy-seven different scales.

Historical temperature records stretch back only 300 years in a few places, but for much of the world it's only been a few decades. The longest series of historical measurements ever assembled was by Professor Gordon Manley, a British scientist. That record gives the mean temperatures for every month from 1659 to the 1970s at a typical lowland site in central England.

Earth's temperature for 140 years.

(Illustration courtesy IPCC)

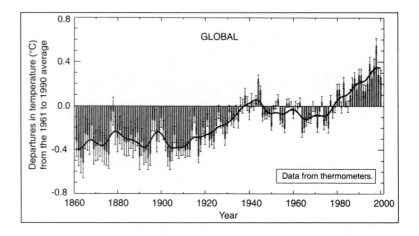

Trying to guess the weather without a thermometer takes a little more detective work. Grain prices tell you about the weather, because the price of grain corresponds to the yield of the latest harvest. Diaries and chronicles help in places but often don't add much detail. A fourteenth-century scroll from the manor of Knightsbridge, London, refers to a great drought in the summer of 1342. Medieval monastic records have various references to great floods, droughts, and shipwrecks. Often the weather is only mentioned in township chronicles as an excuse for spending more on bridge or harbor repairs.

Climatoids

The oldest year-by-year record that comes down to us is the flood levels of the Nile River in Lower Egypt, which gives us an idea of the summer monsoon rains back then. Some records inscribed in stone go back to the first dynasty of the pharaohs, about 3100 B.C.E.

Some of the best-kept records of temperatures have come from ships. It seems sailors have a personal interest in trying to figure out the weather. Early sailing vessels would take the temperature by putting a bucket over the side, pulling up some water, and tossing in a thermometer. As with many early readings, scientists today have to allow for errors. While you're measuring the temperature in a bucket, there is water evaporating from the side of the bucket, which has a small effect on the temperature of the water in the bucket. Scientists studying ship records have had to go back and find samples of these early buckets, study them, and make adjustments to their readings to get accurate measurements.

It's been the same on land. If you put your thermometer in the middle of a town and a city grows up around it, that affects the temperature. It's known as the *urban heat island effect*. Scientists have worked hard to take those effects out. You can't just take a temperature measurement from old sources and one from a new source and compare the results. You must go back and work on the variables that affect accuracy.

Warm Words

The **urban heat island effect** is the effect whereby temperatures taken in urban areas, or cities, are higher than in surrounding rural areas.

Taking the Earth's Temperature

Satellites aid in determining the temperature of the earth today. They do this by measuring the amount of radiation the earth is sending back to space. If you have an electric stove, and you turn it on low, you don't see it too well, but if you turn it on high, it glows bright red. The planet works the same way. The satellite can look down and determine how bright the glow is. The warmer we are, the brighter we glow.

The thermometric record of the earth's temperature has come under fire because skeptics say that scientists have not properly taken into consideration the urban heat island effect. So scientists have had to look elsewhere to determine if, yes, the world is heating.

One thing they look at is glaciers. Glaciers are weird beasts. They surge, their ends wiggle, but if you watch a whole lot of glaciers for a whole lot of years, they respond to temperature more than anything else. And if you look at records of glaciers over the last hundred years, almost every glacier on earth has gotten smaller.

Warm Words

Glaciers are moving masses of ice and snow that accumulate in areas where the rate of snowfall exceeds the rate at which the snow melts.

There is another effect that global warming scientists use to prove their case. If you take a roast out of the freezer and put it into the oven, then the outside will warm up while the inside will lag behind. If you were to drill a hole and put a thermometer in the roast, you'd find the outside is warm and the inside is cold. That tells you the roast is changing from cold to warm. If the roast had been in the oven long enough, the outside and inside would be the same temperature.

It's the same with Earth. If you drill a hole in the earth and you drop a thermometer 100 meters down, the average yearly temperature is a little bit colder than the surface almost every place on earth. This is telling us that it is warming up. The core of the earth is really hot. It's not a natural phenomenon for the earth to be colder deeper down. It has to mean that it was colder in the past and it is warming up.

And the numbers you get from the tree rings, ice cores, thermometers, glaciers, satellites, and from below ground temperatures all say the same thing—the earth is getting warmer.

The Ancient Record

To read the history of the earth's temperatures, scientists look at ocean bottoms, lake bottoms, ice sheets, coral reefs, and tree rings. The layers in these phenomena represent a history book of the past. Sediments in ocean bottoms, lake bottoms, and ice sheets come from snow, rock, plants, pollen, and the skeletons of living organisms accumulated, layer upon layer, over the ages. Corals and tree rings give us a history of the weather and the health of the ecosystem. Scientists looking at these things are doing the same thing archeologists are doing when they peer through trash dumps of ancient cultures. They are looking for clues to the past.

> **Warm Words**
>
> **Sediments** are fragments of inorganic and organic material that are carried and deposited in layers by wind, water, or ice.

Ice Cores

One of the focuses of the International Geophysical Year of 1957-1958 was the little-explored icy regions of Greenland and Antarctica. A number of coring projects started where scientists drilled into the ice and extracted long frozen cores. But the initial cores were taken in less-than-ideal places, and so in 1989, an international consortium was established where representatives from Europe and the United States drilled holes 20 miles apart at the summit of the Greenland ice sheet. Though a portion of the world's permanent ice is locked into *glaciers* in high mountain regions around the world, fully 99 percent of the earth's ice is in the great *ice sheets* that cover most of the island of Greenland and nearly the entire continent of Antarctica.

> **Warm Words**
>
> **Ice sheets** are *huge* glaciers like those that cover Greenland and Antarctica.

Richard Alley, Evan Pugh Professor of Geosciences, was a part of the Greenland Ice Sheet Project 2, the

U.S. contingent of the ice core research. He describes Greenland as a "wonderful place but very cold. The average temperature on the dome was -31°C." The temperature gets up above freezing only about one day per century and scientists can actually see this in the core samples they take from the ice as areas of snow melt.

The midnight sun circled the horizon as the men worked to bring up their samples. Most of the men wore thick, insulated jumpsuits like you might see workers wearing in an ice cream factory. The men stayed in tents or canvas Quonset huts. The scientists worked alternate 6-week shifts through the summer, though the drillers hung around for 12 to 13 weeks.

The ice sheet that lies over central Greenland is an accumulation of compacted snow turned to ice that is about 2 miles thick. About 10 percent of the volume of glacial ice is air, which when compacted gathers into little bubbles that are samples of the atmosphere at approximately the time the ice was formed. Alley studies ice cores from the past to give us clues as to how the temperature is going to change in the future.

> **Global Warnings**
>
> If all the ice on Earth were to melt right now, the global sea level would rise 200 feet. That would be enough to move the coast of Florida somewhere up into Georgia.

To get at those samples of air, he and his associates drilled down into the ice with equipment that looked very much like an oil derrick, including a 100-foot tower, miles of pipe, and a large geodesic dome around the base that kept the weather off the platform.

Teeth were cut on the end of the eight-inch thick pipe, so that as the drill descended it captured a sample core of ice in the middle of the drill pipe. The core samples that came out of the drill were about 20 feet (6 meters) long. Alley and the other scientists took those down to $6^{1}/_{2}$ feet (2 meters) samples for analysis and then 3 feet (1 meter) for shipping. Samples were sent via refrigerated airmail to the National Ice Core Laboratory in Denver, a project supported by the National Science Foundation and the U.S. Geological Survey.

So What Have We Got?

A lot of the analysis, however, was done in Greenland. Scientists took an immense snow blower and cut a 20-foot-deep trench in the snow, put a lid on it, and created a subsurface lab right on the summit of the ice sheet. As the cores were brought into the lab, the scientists made lengthwise slices of the segments to be analyzed by the various chemical, electrical, and visual methods.

When the cuts were made, the different scientific groups split up, taking their samples into little rooms cut off of the trench for further analysis. One of the processes used was to take two electrodes and draw them down both sides of the length of the core. This measured how electricity flowed from one electrode to another through the ice. If there was a lot of volcanic acid in the ice, electricity flowed well. If there was a lot of dust in the ice, electricity didn't flow well at all.

In central Greenland, the snowfall accumulates on the summit and because the temperature only very rarely rises above freezing, the snow doesn't melt, but instead accumulates and is compacted into ice from the weight of the snowfall above. About 12 inches (30 centimeters) of ice represents a year in the top portions of the ice sheet. But as the layer gets buried deeper and deeper in the ice, it gets more compacted. When the layer gets buried halfway through the ice sheet, the layer gets stretched and thinned to 6 inches (15 centimeters) thick. When the layer is buried three quarters of the way through the ice sheet, the layer is only 3 inches (7.5 centimeters) thick. Compaction increases as it goes deeper to the point where it becomes difficult to read.

Climatoids

Greenland contains air bubbles that are samples of the ancient atmosphere. The bubbles are compacted as they sink deeper into the ice and, according to old polar hands, the ice is great for cocktails because of the loud pops the ice makes when you pour alcohol on it.

Warm Words

Proxies are methods of determining values such as temperatures and rainfall by using substitutes, which give indirect measurements. **Isotopes** are varying forms of an element that have closely related properties but different atomic weights.

Though scientists use various chemical and electrical methods to determine the age of the section, Alley and his associates were also able to determine annual layers in the ice by putting them on a light table. The low-density, coarse-grained layers formed by the summer sun show up as light bands, while the fine-grained snow packed by winter storms and winds appear darker. When they checked their estimates of age against the fallout of historically dated volcanoes, which appear in the ice as layers of ash, they found they had it right.

Inside the ice cores are samples of the air taken from near the time when the snow first fell. The air in the ice collects into bubbles and is further compacted, but upon release is a fairly accurate representation of the atmosphere at approximately the time the ice was formed.

Scientists can't take measurements of the rainfall and temperature in core samples, so they use *proxies* to help them out. Determining the varying amounts of oxygen *isotopes* is one proxy method that tells scientists the temperature of the atmosphere when the rain fell.

One of the isotopes that scientists look for in Greenland ice is oxygen-18. Regular oxygen has 8 protons, and 8, 9, or 10 neutrons, which combine to make oxygen-16, oxygen-17, or oxygen-18. When oxygen-18 combines with hydrogen it forms heavier water. Though it constitutes less than 1 percent of the water, all water samples have a little heavier water and it can be measured.

When a raindrop forms in a cloud, the cloud feeds water to the droplet, and because oxygen-18 is heavier, it falls out of the clouds faster. Scientists find a higher ratio of oxygen 18 in the raindrops and a lower ratio in the clouds.

Because cold air holds a lot less water vapor than warm air, clouds moving from the tropics toward the poles lose water all the time. As the clouds lose more water they deplete their stores of oxygen-18. The further north the cloud goes, the more it gets depleted in oxygen-18. It turns out that if you measure the amount of oxygen-18 in snow or water, it is indirectly proportional to the temperature when it fell. Scientists can use oxygen-18 content to determine the temperature when the snow fell in an ice core, or when a tree took up water, or when shellfish incorporated the water into its shell.

A Varied Record

Greenland is an excellent place to study ice since it collects evidence of the earth's atmosphere from halfway around the world. It gathers snowfall from Greenland. It collects dust blown to it all the way from Asia, sea salt from the ocean, and methane that comes to it from swamps in the tropics.

Methane is among the many gases that interest Greenland ice core scientists. Methane prior to man taking over the earth was essentially swamp gas. Huge changes in swamp gas tend to go right along with changes in dust and temperature. If there is a lot of methane in bubbles from gas core samples, it means that at the time those bubbles were made, there were a lot of swamps in the world. Determining whether those swamps were in the tropics or in Siberia is another part of the riddle.

To solve this riddle, scientists also measure the methane in air trapped in ice cores in Antarctica. Because methane has a limited life

Climatoids _____

The winds of the world do a good job of mixing the atmosphere. If you were to release a large quantity of gas in Seattle, it would only take a few years before people in Bangkok could measure what you released. Dust and sea salt don't mix as well, but most gases are in the atmosphere long enough to mix globally.

span in the atmosphere, the further it gets from the source, the less methane will be in the atmosphere. So to determine where the methane in the ice is from, you have only to compare it with the methane in the ice from Antarctica. If the methane is from the tropics near the equator, then there will be an equal amount in both Greenland and Antarctic ice. If the methane is from Siberia, there will be more methane in the Greenland ice—because Greenland is closer to Siberia—and less methane in the Antarctic ice. Most methane in the ice is from the tropics, and its presence tells scientists how extensive the tropics were during the period the bubbles represent.

Carbon Dioxide in the Ice

By having samples of the ancient atmosphere, scientists at Greenland have been able to measure the amounts of carbon dioxide in the atmosphere in historical times. Since carbon dioxide is the main greenhouse gas responsible for global warming, scientists have looked in the ice cores to investigate its history.

What they've found is that over the last 110,000 years, the concentration of carbon dioxide has never been nearly as high as it is today. In 1750, at the beginning of the Industrial Revolution, the concentration was about 280 ppm. The present concentration is about 367 ppm and rising, an increase of about 31 percent, and an 18 percent jump in just the last 40 years.

That rise is unprecedented in the last 10,000 years. If you go back before 1850, the concentration of carbon dioxide in the atmosphere has remained steady over time. It's always been about 280 ppm. It hasn't varied in the last 10,000 years. This fact adds credence to the assumption that the current rise in carbon dioxide is derived from human causes, since 1750 is around the time that people started burning lots of fossil fuels.

If you go back even further, before 10,000 years ago, there are some variations. During the last ice age carbon dioxide concentrations were about 200 ppm. It varied with the ice ages as they came and went approximately every 100,000 years. During the last four glacial cycles over the last 400,000 years, carbon dioxide has gone from 280 to 200 ppm when it got cold and then back to 280 ppm during the interglacial warm periods. It never got above 280 ppm. Today's concentration is unprecedented.

> **Hot Debates**
>
> Scientists believe that the decrease in the level of carbon dioxide from 280 to 200 ppm probably amplified the ice ages, though it didn't cause them. The decrease was caused by the earth's orbit around the sun. But the orbit can't explain all of the climate change. The carbon dioxide change amplified those changes. During the ice age it made the cold parts much colder.

Why the carbon dioxide dropped during the ice ages is a mystery. Some scientists think it has something to do with a reorganization of how the oceans circulate, particularly around Antarctica where the deep ocean comes up to contact the surface, and surface waters descend to take their place. It's thought that if you change this circulation, then one of the results might be that the plants in the ocean would do a more efficient job of removing carbon dioxide. It's something called the *biological pump*, where if you change the way the waters circulate you may increase the amount of carbon dioxide taken from surface waters.

Warm Words

The **biological pump** is a mechanism in the sea whereby plants such as algae remove the carbon dioxide from the surface. Animals then eat the algae, and the carbon dioxide ends up in their feces, which settle to the bottom of the sea. That makes room for more carbon dioxide at the surface.

Analyzing the Climate Clues

Greenland ice is not the only method scientists use to study ancient climates. There are a number of others, including tree rings, lake bottoms, swamp bottoms, sea bottoms, and coral samples that give scientists a look into the past. Scientists use measurements from these sources to provide additional information and also to double check original sources.

Tree Rings

Have you ever counted the annual rings of a tree stump to figure out how old it was? The thickness of the tree rings tells scientists something about the climate and how happy the tree is with climate change. If it's in a place that's cold, it will tell you it's happy when it's a little warmer. And if a tree is in a place that's really dry, it's happy when it's a little wetter. ("Happy" equals a thicker tree ring.)

Fortunately we don't have to cut the tree down to get at the climate record in its growth rings. Scientists can extract this record by taking a pencil-thick core from the tree without stunting the tree's growth. Ring widths in trees that occur at the upper altitudes or poleward limit of the

Climatoids

At the Laboratory of Tree Ring Research of the University of Arizona in Tucson, scientists have built a chronology of ring widths from bristle cone pine trees in the White Mountains on the California-Nevada border. Those trees are some of the oldest trees on Earth and University of Arizona scientists have constructed a tree ring record that goes back almost 9,000 years to 7,000 B.C.E.

tree's range give scientists the best picture of the climate. If the ring is thick, it probably indicates a warm year. If it's thin, it indicates a cool summer. In trees that are found in the warm arid margins of their natural extension, a good ring could indicate good rain, whereas a narrow ring could indicate drought.

Lakes and Peat Bogs

Sediments in lakes and peat bogs give scientists their best clues as to the plant vegetation in a given area. The layers cough up pollen samples or beautiful fossilized leaves of ancient plants. These give scientists clues to the atmosphere because they know what climate the plants prefer.

One of the first histories of the post-glacial renewal of vegetation was written in 1876 by Norwegian botanist Axel Blytt. In his *Essay on the Immigration of Norwegian Flora* he detailed from bogs and sediments how the tundra established itself in the path of the retreating glaciers, later to be replaced by birch, then birch and pine, and later by broadleaf trees.

Global Warnings

How much would El Niño change in a warmer or colder climate? Interest in that question has recently sharpened because there are indications that the frequency and intensity of El Niños underwent a shift beginning in the 1970s. Some scientists believe that the tropical Pacific might be the driver that caused the climate to flip between interglacial and glacial periods.

Corals

Corals provide a good indication of the annual cycles of El Niños in the southern hemisphere. El Niño is important because it is the largest recurring climate event in the world, outside of the seasons. Changes in El Niño weather have the ability to drag along the weather of the entire tropical region and thus are considered a natural candidate for a globalizer of climate influences.

Furthermore, comparisons with prehistoric El Niño records recovered from corals suggest a systematic relationship between global conditions and the character and strength of El Niños.

Sea Bottom

There are a number of things one can learn from the bottom of the sea. For example, marine plants incorporate stiffer molecules in their cell walls to offset the softening effects of higher temperatures, and the fraction of stiffer molecules in sediments yields an estimate of past temperatures.

Scientists at the University of California at Santa Barbara were able to put together a record of the climate history of the earth going back over the last five ice ages from the ocean bottom in the tropical Pacific. They discovered that forams, a tiny sea creature, have shells that differ in their magnesium content depending upon the temperature of the water.

Most ocean bed deposits accumulate extremely slowly, $1/4$ to $1^1/_2$ inches (1 to 4 centimeters) per century is typical. They incorporate not only the remains of marine organisms but also mineral dust blown by the winds. Though they are not detailed indicators of relatively small periods of time, they can yield extremely long records.

Climatoids

The longest sedimentary record so far is from the equatorial Pacific and was analyzed by Cambridge University scientists. It goes back more than two million years and indicates that the ice ages have occurred approximately every hundred thousand years during the last million years or more.

Atomic Subs

The Arctic has warmed markedly in the last two decades, and scientists really don't know why. Knowledge of the Arctic lags behind that of the world's other oceans. Covered by ice year-round, the Arctic is virtually inaccessible to ship traffic and has been the province largely of nuclear subs trying to hide from one another on under-ice missions during the cold war.

Now, however, some of those subs are taking scientists along for their first sub-Arctic views as part of the SCICEX (Science Ice Expeditions) program, spearheaded by the U.S. Navy and the National Science Foundation. Their discovery is that the ocean's ice cover has thinned by an average of 4 feet (1.2 meters)—some 40 percent—since the 1960s. Some scientists think that the ocean's summer ice-over could completely stop at some point in the coming decades. The Arctic Ocean may have lost its summer ice 400,000 years ago when the earth was as warm as it is now.

The Least You Need to Know

- We've been taking the temperature of some parts of the world since Galileo.

- To get a picture of the climate before that requires us to look at glacial ice; tree rings; sea bottom, coral, and lake sediments.

- Scientists taking ice cores in Greenland have a 110,000-year record of the climate, including direct samples of ancient air.

- All evidence leads to the conclusion that the world is getting warmer and carbon dioxide concentrations are increasing in the atmosphere.

Chapter 6

Climate Is a Roller Coaster

In This Chapter

- ◆ The climate can change abruptly
- ◆ Changes in the last ice age
- ◆ Changes closer to the present
- ◆ Breaching climate thresholds

In the previous chapter, I talked about how scientists have collected evidence of the climate from the past—digging in ice, stomping around in old lakebeds, and bringing up samples from the very depths of the sea. Much of the new evidence has only been gathered since the International Geophysical Year in 1957-1958. But most of the big push has occurred in the last two decades. In looking at the past, scientists have paid particular attention to abrupt changes in climate and how they have affected our planet Earth. Society is used to gradual changes. But what would happen if different climate regimes came more suddenly? What if you woke up one winter day in New York City and the weather was like Miami? And what if it stayed that way? How would that affect life, the environment, and the economy?

We're Spoiled

Face it, we're spoiled. We take our fair weather for granted. The rain we have, the storms we get, the droughts we endure seem severe. One summer is cool, the next is hot. One winter it snows a lot, the next there's hardly any snow at all. The weather is wild. It must be changing. Didn't I hear something about global warming? Is that what's going on?

It's true we are under the influence of global warming, but perhaps the reason we're so complacent about it is we're spoiled. The weather we've endured—that flood where your car got stuck, the fire that stopped your vacation, the hurricane that blew the windows out of your Aunt Myrtle's summer home in Florida—is, I'm afraid, peanuts compared to the climate records that scientists have uncovered of days gone by.

The earth right now is in an *interglacial*. It's the period between glacial episodes. The world is warmer, the winds don't kick Asian dust and sea salt in your face. Those glaciers that once hung outside Chicago and New York are gone. When I step out my door in southern California, the climate isn't like Vancouver one year and Acapulco the next. But according to ice cores, tree rings, and sediments, such climate swings may once have been the norm.

Warm Words

The **interglacial** is the period of relatively stable climate between glacial episodes. The earth is currently in an interglacial.

Ice age glaciers carved this lake basin.

(Photo by Michael Tennesen)

The Ice Ages

As I mentioned in Chapter 5, the longest continuous sedimentary record to date goes back more than two million years and indicates that glacial episodes have occurred approximately every hundred thousand years during the two-million-year-old ice age we are in right now. To get an idea of what it was like before that, scientists look in Earth strata for *tillite*, deposits of rock and clay that tells them there's been a glacier here. From this evidence, scientists believe there have been at least six ice ages in the past, periods when ice covered extensive areas of the earth. The last ice age was from 270 to 350 million years ago. It may have had more than 30 glacial epochs.

> **Warm Words**
>
> **Tillite** is a layer of rock and clay that are evidence of glacial deposits.

Prior to the current ice age, we were in a hot-house climate where there was little or no permanent polar ice. Scientists believe this warm period extends back at least 65 million years. Presently about 10 percent of the earth's water is locked up in ice, mostly in the Greenland and Antarctic Ice Sheets. About 20,000 years ago, about 30 percent of the world's water was locked up in ice, and Greenland temperatures were about 20°F (11°C) cooler. The coldest parts of the ice age might have been as much as 40°F (22°C) cooler.

> **Climatoids**
>
> Glacial deposits from the last great ice age, 270 to 350 million years ago, are found in South America, southern Africa, and Australia. German meteorologist Alfred Wegener in 1912 used those deposits as evidence of his theory of continental drift, which held that all three of these countries were once a single landmass known as Gondwanaland near the South Pole.

Peaks and Valleys

It's hard to talk about average temperatures when describing the most recent glacial episode. The climate didn't just get cold and stay that way, it swung wildly back and forth. Over the last 10,000 years or so, the temperature has stayed relatively warm and constant. But about 12,800 years ago when the earth was coming out of the last ice age and things were warming up nicely, the earth suddenly took a big dive back into the cold. Then a little later it jumped back into the warmer period. It was called the Younger Dryas event. I'll get into it a little later in this chapter.

The Younger Dryas event might just be business as usual so far as the climate was concerned. The Greenland record shows that during the last glacial episode large

Global Warnings

Some say the world will end in fire,
Some say in ice.
From what I've tasted of desire
I hold with those who favor fire.
But if it had to perish twice,
I think I know enough of hate
To say that for destruction ice
Is also great
And would suffice.

—Robert Frost, "Fire and Ice" (1923)

abrupt climate changes have occurred 24 times. Evidence from the glaciers, salt from the ocean, methane from the tropics, wind-blown dust from Asia, and snow from Greenland, also indicates that those big climate swings were world-wide.

Over the last 100,000 years there have only been two vaguely stable periods in the world's climate. One was a short period back in the middle of the last ice age when the weather was at its coldest. Central Greenland was 40°F (22°C) colder than what it is today. (Imagine what it would be like if the average temperature of the place you call home dropped that much.) The other period of stability is the last 10,000 years. The rest of the time, it's the climate roller coaster. Some of these dives and jumps took less than a decade, perhaps a few years.

Solar Wobbles and Wiggles

So how do we get the glacial episodes of our present ice age? Why does the world seem to be on a 100,000-year cycle of glacial ice? (Remember, if we can figure out the past, maybe we can figure out the future.) The earth rotates around the sun, but the path it takes is not a perfect circle. The orbit of our planet is out of round; they call it "elliptical". Sort of like a football, only not so pointed. But on a 100,000-year cycle the orbit changes. The orbit becomes less like a football, and more like a basketball. And this affects the climate. Note that the 100-year cycle matches the length of the glacial episodes.

Hot Debates

Scientists believe that we are currently about 11,500 years into in an interglacial, the warm stable period between two glacial episodes of the present ice age. Interglacials are typically 10,000 or 20,000 years, depending upon which scientist you're talking to. Which means we are either out of time and about to fall off into the cold abyss, or we might have another 10,000 years to go.

There is another variable. The earth's axis is on a tilt. As it travels around the sun, it's a little cockeyed, like your globe is. This is what makes the seasons. The northern hemisphere of the globe is tipped toward the sun on one side of the orbit, which is when the northern hemisphere has summer. The northern hemisphere is tipped away from the sun, on the other side of the orbit, which is when it has winter. The axis also wobbles over time.

Scientists have long looked to the eccentricity of the orbit and the tilt and wobble in the axis as the main reasons for the glacial episodes in our current ice age. They don't increase or decrease the total amount of sun the planet gets, but they change where and during what season the sun falls, and this is enough to cause the glacial episodes— or so we used to think. However, we are now in the age of computer models. We can take certain climate variables, feed the data into a supercomputer, and get a prediction. We can also do it the other way around; feed the result into the computer and see if the variables we are investigating give us the result that we already know. And the problem is if we feed the eccentricity of the earth's orbit and the tilt and the wobble of its axis into a computer, they are not enough to produce the glacial episodes in the current ice age. Maybe it helped get things going in the right direction, but it didn't pull off the freeze all by itself. So what did?

Who Done It?

Could it be the ice itself? Ice has a greater *albedo*. It reflects sun more than the forest and more than the dark seas, which absorb the sun and heat the place up. Ice sends sun packing. If you get more ice, you get more reflection, so it could perpetuate itself. But ice does some funny things that keep it from taking over. As the ice sheets grow bigger, it acts as a blanket, warming up the earth beneath it. Plus, the ice sheet is so heavy it sinks the earth beneath it to altitudes low enough to raise the temperature. When it gets warm down there, the ice can lose its frozen grip on the land, and slide out to sea. Or it can simply drown in a pool of warm water. In other words, glacial ice normally has a few governors, which keep it from becoming self-perpetuating.

Warm Words

The **albedo** of a surface is its ability to reflect back the light that falls upon it.

So what else? Well, how about carbon dioxide? As it turns out, the same stuff we're worried is going to make us warmer also had a major role in making the planet Earth a lot colder. Carbon dioxide in the atmosphere, deep-sea temperatures, and the orbital eccentricity of the planet's orbit all vary in step with the same 100,000-year cycle as the ice ages. Ice volume lags behind a bit, apparently ruling it out as a prime mover. Deep-sea temperatures may have something to do with the conveyor-belt-like currents that are operative over most of the oceans, moving warm water and cold water at different levels. A breakdown in this system may be a part of global warming. But I'll get to that more in Chapter 13.

But here's our old friend carbon dioxide. For reasons that scientists are still not clear about, it seems that carbon dioxide decreases with the beginning of the ice ages and increases on the way out. Scientists have seen this in the Greenland ice cores, but also in the much longer records of ice cores in the Antarctic, which goes back over 400,000 years. Richard Alley feels that for reasons we don't understand completely, the cooling of the ice-age cycles driven by the eccentricity of the earth's orbit reduces the greenhouse gases, particularly carbon dioxide and water vapor. Water vapor fell when ice-age cooling reduced evaporation. Carbon dioxide fell, he believes, when blustery ice-age winds blew more dust into the oceans, fertilizing the algae. Algae (or phytoplankton) is part of the biological pump, you remember, whereby plants suck up the carbon dioxide, animals eat the plants, and the carbon dioxide ends up as old poop or old shells at the bottom of the sea. Colder seas also suck up more carbon dioxide even without the little plants and animals.

> **Climatoids**
>
> Colder seawaters hold carbon dioxide better. It's the same as in a can of soda pop. If you heat it up, it loses its fizz, which is the carbon dioxide. Conversely, if you cool the water down, it absorbs more carbon dioxide. The problem is, cooling your soda down won't put back the fizz if you've already opened the can. The carbon dioxide is gone.

Snowball Earth

Sure, the last glacial period of our present ice age was rough, but the saber-toothed tigers, the wooly mammoths, and ice age man, who used to complain about it so much, didn't know how good they had it. What if they'd been around during snowball Earth? The present ice age is peanuts compared to what went on 600 million years ago, during a period known as the Neoproterozoic, when some scientists believe that the surface of the planet almost froze solid.

Scientists got their first hint of snowball earth from glacial debris found near sea level in the tropics. The thing is, most glaciers near the equator never get down below 16,400 feet (5,000 meters). Even in the coldest winter of the last glacial period they never went lower than 13,100 feet (4,000 meters). If scientists who studied snowball Earth can be believed, the worldwide global temperature got down to an

> **Global Warnings**
>
> Life got clobbered during snowball Earth. We're not talking guys and girls in business suits or even a good brontosaurus. Life in the Neoproterozoic was something you'd look at through a microscope. But carbon deposits in the rocks hint at a dramatic loss of life during this event. Only a few microscopic organisms survived by huddling around volcanic hot springs.

average of -58°F (-50°C). Heat, ultimately generated by the hot core at the center of the earth, kept the ocean from freezing to the bottom, but it got so cold that sea ice froze to 6/10 of a mile (1 kilometer) in thickness. Microscopic life on the planet was nearly wiped out.

Scientists believe that torrential rain (which binds to carbon dioxide) removed carbon dioxide from the atmosphere, during a period when most of the continents were gathered around the tropics. This and the albedo of polar ice caps led to a runaway freeze.

Hothouse Earth

But then what? Well, after snowball Earth we had hothouse Earth. Even though the earth was frozen, volcanoes spewed a steady stream of carbon dioxide into the atmosphere. Carbon dioxide levels increased about 1,000-fold over 10 million years of normal volcanic activity. The ice melted and the runaway hothouse formed. In a matter of a few centuries, a brutally hot world replaced the former deep freeze. Surface temperatures soared to more than 122°F (50°C), driving an intense cycle of evaporation and rainfall.

Climatoids

Only a tiny fraction of life survived snowball Earth. In terms of biological destruction, it was more deadly than the period following the asteroid hit that took the dinosaurs out 65 million years ago. Then came hothouse Earth, and the scorching heat further thinned the small numbers of organisms that survived the ice. But then Earth settled back to normal. A rapid diversification of multi-cellular life followed, resulting in the Cambrian explosion of organisms between 575 and 525 million years ago. It seems hard times promotes genetic diversification. Or better put, only the strong survive.

What's impressive in many of these great catastrophic swings in the climate is the role that carbon dioxide plays. It is one of the main drivers of the climate on the planet Earth. Swings in its concentration have molded the land and left massive scars on the terrain, as well as on life itself. The more scientists study its effects, the more they realize it's not something to play around with.

Suddenly it got very cold.

(Photo by Michael Tennesen)

Meet the Younger Dryas

But, you say, this was all a long time ago. Most of this stuff took eons. The geological history of our planet is on a clock that moves over such vast amounts of time that it's not applicable to human life. Why worry about something that moves so slowly that it can't possibly affect you in your lifetime or your children's lifetime or even your grandchildren's lifetime?

Well, let me introduce you to Younger Dryas. The event occurred about 12,800 years ago during a period of increasing warmth, when the world was suddenly and very drastically thrown back into the last glacial period of the present ice age. It remained that way for more than a millennium, until about 11,500 years ago when it swung back out. The warming on the way out of this event was very sudden. Indications are that the average Greenland temperatures at the end of the Younger Dryas event jumped 15°F (8.4°C). According to a number of different studies, the jump happened in decades or less.

CAUTION

Global Warnings

The data indicates that the cooling into the Younger Dryas event occurred in a few decade-long steps, whereas the warming at the end of this enormous climate swing occurred primarily in one especially large step of about 14°F (8°C) in around 10 years. That jump was accompanied by a doubling of snow accumulation in 3 years, most of the change occurring in one year. The Younger Dryas event is a term that refers to this whole period.

Scientists digging into peat bog sediments in Europe in the late 1800s were the first to discover it. The *dryas* is an arctic flower in the rose family that is white with a yellow center. It's found in cold tundra or in high altitudes. Digging through sediments to determine the sequence of plant types following the retreat of the glaciers, scientists found a number of layers representing different forest types that occurred as the earth warmed up. But then they found the small little fossil leaves of the dryas. Going deeper they encountered three layers of these plant sediments before encountering the rock and gravel that were laid down by the glaciers themselves. Those layers represented the Younger Dryas, the Older Dryas, and the Oldest Dryas. The Younger Dryas was the most severe. It's also the most studied of any abrupt climate change on our planet.

Warm Words

The **dryas** is an arctic white flower with a yellow center, a member of the rose family, found in the tundra. It's used as an indicator of arctic conditions in sedimentary layers.

Earth's Climate Versus Younger Dryas

Ice core evidence from Greenland shows that wind-blown dust and sea salt increased by a factor of 3 to 7 times. In Norway, mean July temperatures were about 12 to 16°F (7 to 9°C) colder than today. In Spain, the July temperatures were as much as 14°F (8°C) colder than today. For a while, scientists thought the event was mostly confined to Europe, being the result of a shutdown in deep north Atlantic currents. But Dorothy Peteet, a senior research scientist at NASA in New York, found evidence of Younger Dryas in lakebed sediments in the Adirondack Mountains and on the islands off Alaska.

She found that the tundra moved all the way to Virginia, replacing the forests. When she'd gotten tired of looking at Younger Dryas in the Northeast, she went on an expedition to Alaska where she discovered lakebed sediments of fern overlaid by tundra plants on Kodiak Island. "We joked with each other that it was probably Younger Dryas, and it turned out that was exactly what it was."

Younger Dryas evidence also showed up in Central America, the Caribbean, and South America. Marine sediment cores from the event show an increased abundance of polar plankton. This hints that deep ocean currents might have weakened or shut down during the Younger Dryas, and may even be the reason for it.

But the greatest catastrophe of the Younger Dryas event according to Dorothy Peteet was the extinction of the great animals of the present ice age. The mastodons, mammoths, horses, and saber-toothed tigers, which once wandered the North American continent, became extinct during the Younger Dryas.

Hot Debates
Scientists debate whether the extinction of the mastodons, mammoths, horses, and saber-toothed tigers, which occurred during the Younger Dryas (12,800 to 11,500 years ago), was the result of early man's invasion of North America or the massive swing in climate during that period. Many scientists believe that both of these causes contributed to one of the highest periods of mass extinction in North America.

What If It Happened Today?

According to Peteet, if an event on the scale of the Younger Dryas warming (the period coming out of its cooling) was to occur in the Northeast, it would be catastrophic. "The weather in New York and New England would suddenly become like Florida. We would lose all the trees in the Adirondack Mountains—the spruce, fir, and paper birch. We'd also lose the oaks and white pine. The dead wood would create major forest fires. And one tree from Florida wouldn't move up to replace them. We don't have migratory pathways. We have highways and agriculture instead."

Rapid changes such as the Younger Dryas are basically harder for man and nature to endure. While the economy, industry, and man may be able to replace, repair, and adapt to changes that occur over time, rapid change is tough. A rapid sea level rise, for example, would devastate coastal buildings and harbors.

Rapid warming could destroy forests, vineyards, and fruit trees, not to mention the ski industry. It could rapidly deplete the ground water reserves of the Southwest desert. You could expect your property value to fall if your place wasn't insulated or cooled properly. Droughts would devastate native trees, which need water to fight off insects.

The Mechanics of Abrupt Change

So what do we mean by an abrupt climate change? Twenty-first century man thinks of the climate as something, which, though it may vary from year to year, stays pretty constant. You can rely on the Northeast for cold, the Northwest for rain, the Southwest for sun, and the Southeast for subtropical storms and sun. But that's not really the way the climate works. That's just our recent experiences. The way the climate really works on our planet is that it jumps all over the place.

The analogy of abrupt climate change is shown in the previous figure. Imagine a balance consisting of a curved track where there are two cups in which a ball might rest. The system has three points of equilibrium: (a), (b), and (c). The middle point (b) is unstable. If the ball is moved ever so slightly to the left or right, the ball will move to a point far from its original position. However, if you were to move (a) or (c) a bit, the ball would merely rock back and forth and then settle back into the (a) or (b) cup. But if you push down on either (a) or (c) gradually, more and more, eventually the ball will roll over the fulcrum and pass into the opposite cup. This is an example of a system passing a threshold. When the pressure is relieved, the system does not return to its former state. It has entered a new regime.

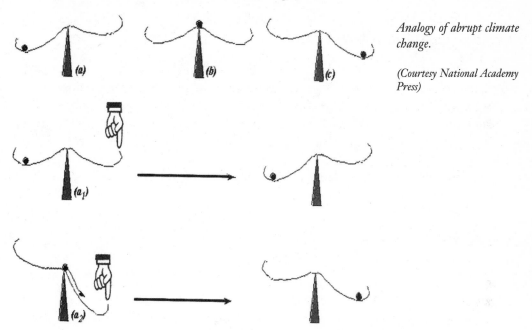

Analogy of abrupt climate change.

(Courtesy National Academy Press)

This is what scientists feel happened with Younger Dryas. There was steady force exhibited by various climate features, until the climate shifted into an entirely new regime. The climate over most of the world changed suddenly and drastically.

A good example of a system passing a threshold and entering a new regime is the floods on the Red River in Grand Rapids, North Dakota in the late 1990s. Since the nineteenth century, Grand Rapids had successfully fought frequent floods on the Red River to a river stage of 49 feet. Then in 1997, the water level crested the levees at 54 feet, and led to a catastrophic flood. You might say that the threshold had been passed, after which the levees failed. When the city rebuilt the levees, they considered past events to determine what was the highest that the river might rise.

The trouble is the world might be moving into a new era, where the thresholds for both individual systems like the Red River as well as large systems like the deep north

Atlantic currents may well be breached. The constant pressure that we are putting on this system by a gradual rise in carbon dioxide may lead to a corresponding gradual rise in temperature, or it may lead to a sudden, widespread, and catastrophic change.

And it just might stay that way for 1,000 years.

The Least You Need to Know

♦ The climate over the last 100,000 years can best be described as a roller coaster with lots of steep drops and rises.

♦ We are currently in a period of calm following the last ice age.

♦ That period is not typical of the earth's climate. The climate in the past has swung from a solid ball of ice to a scorching hot house.

♦ The Younger Dryas event was a dramatic swing in and out of a cold spell in the climate that occurred from 12,800 to 11,500 years ago.

♦ Our climate is like a balance in which the ball could roll from one side of the fulcrum to the other and not go back for 1,000 years.

Chapter 7

More Recent Changes

In This Chapter

- ◆ Ice age hunters start growing stuff
- ◆ The weather and early civilizations
- ◆ A four-century dip back into the cold
- ◆ The dust bowl

In the previous chapter, we described the harsh environment that existed before the Holocene, the relatively warm and stable 10,000-year period we are living in today. Though modern man was around for at least 30,000 years before the end of the last ice-age cycle, it was only when the climate settled down and things got warmer, more stable, and less stormy that civilizations started to flourish.

Early Man Warms Up

The earliest historical records that ancient humans left us are from sketches and paintings on the walls of caves in France and Spain which scientists date back 15,000 years. Those pictures show a world of horses, deer, wild cattle, rhinoceroses, and mammoths which early hunters chased down with spears and arrows. The pictures depict a treeless landscape, where scientists tell us that cold dry winds, storms, and widely varying temperatures dominated.

Those Burly Hunters

Man populated areas that were generally beyond the range of the great ice sheets. Scientists have uncovered a number of their remains, which show the hardy life these guys endured. An ancient encampment in northeast Russia is filled with the bones of mammoths, which means that they weren't just dreaming about the big beasts, they were actually living on them.

Archeological finds show human beings lived in the ice-free areas of Alaska north of the big ice sheets. Humans probably roamed back and forth from Siberia to Alaska across the dry plain that then linked those lands.

Climatoids

During the last ice age, the expanded glaciers and ice sheets locked up enough water to lower sea level around the world by about 400 feet (125 meters). This was apparently enough to produce the Bering Land Bridge from Siberia to Alaska, across which man first ventured into the Americas. Similar great stretches of dry land might have also existed between Asia and Australia, providing the first path man took to the land down under.

But about 15,000 years ago, the world started to defrost. The glaciers receded, vast sections of the ice caps melted, and the seas began to rise. As the world warmed up, more water was available to the atmosphere and rains increased. The great ice sheets that covered Scandinavia began to melt about 8200 B.C.E. and were about the size they are today around 6000 B.C.E. The sea invaded the Hudson Bay about 6,000 B.C.E. and the North American ice cap, which covered much of Canada, rapidly melted.

The sea level rose about 10 to 16 feet (1 to 5 meters) per century. Though the numbers don't seem high, modern experience with sea level rises on the low-lying coasts of the North Sea shows that the coastal land doesn't give way gradually but in sudden retreats during great storms that arrive as storm surges at low tides.

Settling Down on the Farm

Early man lived in mountain caves close to areas where the game was plentiful. However, about 10,000 B.C.E. the human inhabitants in the Zagros Mountains in the Middle East began to change. Understanding their story involves picking through their remains. Up until then, they'd hunted mostly wild sheep and goats. But then sheep became more important to their diet, and the proportion of young sheep bones indicate that they must have started herding.

A large number of grinding tools soon appears, as man began to collect grains. And a little later wheat and barley are found in the steppes near the mountains in areas where neither the climate nor the terrain is suited to the wild forms of these grains. From this evidence, scientists conclude that man must have discovered agriculture.

As the world warmed it got drier. Cave paintings from the Sahara Desert around 3500 B.C.E. show men hunting a hippopotamus from a canoe. Flood records from the Nile River show an increasingly dry world where elephants, giraffes, and rhinoceroses of Egypt disappeared from 3000 to 2000 B.C.E.

The Peak of the Holocene

The Stone Age evolved into the Bronze Age, and the warm weather and freedom from storminess allowed for the development of culture through trade over land and sea.

In the southwestern portion of the United States, the first agriculture spread northwards with increased subtropical rains in the warm period between 4500 and 4000 B.C.E. Native American societies began to flourish.

Tree ring data taken from the bristlecone pines in the White Mountains of California give us good evidence of the history of warmth and rain. That area has gradually dried up and desert has taken over what was once trees and grasslands. A series of lakes that once pooled in Owens Valley between the White Mountains and the Sierra Nevada Mountains today are mostly dry.

Climatoids

The rise of civilization can be traced to that moment when early man left the caves and the more abundant game of the mountains and ventured into open sites in the foothills, which were more conducive to agriculture. This transition took place only after the world began to warm.

Hot Debates

Some scientists believe that the climax of our current interglacial might have occurred around 2000 B.C.E. when the tree line of many of the earth's forests reached their upper limit on many of the mountains of the world. Subtropical and tropical latitudes had their highest levels of moisture and middle and high latitudes their greatest levels of heat.

Religion and Weather

The climate has varied throughout the world. In China there were severe droughts that occurred around 1200 to 800 B.C.E. and again from 600 to 200 B.C.E. Civilization was on the move at that time, perhaps hunting for wetter lands. The movement might have had something to do with the spread of the great religions. Drought brings friction as society struggles for limited resources. It also brings suffering. Buddah and Confucius offered solutions to human suffering and their messages traveled rapidly around 500 B.C.E. This was also the time of the great Greek philosophers whose teachings widely influenced European thought and the rise of Christianity. Christ, however, seemed to have been born in a period of climate stability.

Climate and Rome

The climate wasn't a great obstacle to the Mediterranean world in the time of the Romans. Julius Caesar apparently had to wait in the summer of 54 B.C.E. while persistent winds delayed his expedition across the channel to England. Similar weather could be found today.

> ### Climatoids
>
> Late in the first century C.E., Roman Emperor Domitian prohibited vineyards in the western and northern provinces of the empire. The climate was gentler at the time, and farmers quickly realized that grapes could be grown in places where they hadn't existed before. It was around this time that vine growing began in England and Germany. The edict was revoked by Probus around 280 C.E.; he obviously liked his vino a little more than Domitian.

Greece and northern Africa were probably helped by the colder climates, and this may have been the reason that Roman agriculture proliferated there in that era. The water table, too, might have been higher then and desert oases more extensive, as a result of moister climates that existed coming into that era.

> ### Global Warnings
>
> Though the plague might be most famous for its European invasion in the fourteenth century, it took a toll in Asia in much more recent times. An outbreak in China began in the 1890s and spread over the next 20 years throughout China and beyond, killing more than 10 million. What outbreak awaits a warmer world?

The Big Plagues

A series of plague epidemics swept through the empire in late Roman times. In the second century the plague struck twice in Egypt, reducing the population in various regions by a third. In the middle of that same century, the plague moved from Macedonia and spread throughout much of the empire. The plague struck again in the third century. Worst was the bubonic plague, which arrived from the Middle East in the sixth century. It took a hundred million people in all. The drought that was prevalent in those days may have contributed by making hygiene difficult due to the lack of water.

That's different from the incidence of the disease in the fourteenth century where the plague seemed to spring from warm, moist places in cities. Plague is an infectious fever, a bacterium transmitted by the rat flea. Scientists estimate that the Black Death, as it was known in the fourteenth century, killed more than one fourth of the population of Europe, or 25 million people.

Fall of the Maya

Meanwhile, on the other side of the Atlantic, the Mayan civilization was flourishing in the tropical rainforest of the Yucatan Peninsula in southeastern Mexico. Though the Mayan civilization has lasted through to this day, it reached its apex, so far as control of the population and growth of its great cities, between 300 to 800 C.E.

Mayan ruins at Tikal, Guatemala.

(Photo by Michael Tennesen)

The Mayans spread their influence throughout Central America and Mexico, flourishing in southern Mexico, Guatemala, Belize, El Salvador, and Honduras. Massive temples and stone monuments typify the classic period of the Maya. They had an accurate calendar and even made astronomical observations. They had beautiful multicolored pottery and finely carved statuary.

The population centered in the lowlands of northern Guatemala. Perhaps their best-known site is at Tikal. The center of Tikal contains more than 300 major structures. Temples I and II dominate the central plaza, standing 144 feet (44 meters) and 128 feet (39 meters) high. The adjacent ruins contain palaces and a ball court. Archeologists feel that about 10,000 people lived in an area around the center of the site, while perhaps 40,000 more lived in the surrounding rural zone.

Climatoids

Though Tikal National Park in Guatemala is best known for its Mayan ruins, it's also famous as a wildlife sanctuary. On a visit to Tikal a few years ago, I remember having to stop and wait while a troop of 50 or more coatimundis (raccoon-like animals) crossed our jungle path. I remember walking down lush green trails with ferns, bromeliads, and orchids hanging from every bough, while the spider monkeys crawled through the branches overhead. And I remember spotting a pair of toucans from one of the temples and then following their huge multi-colored bills into the forest. The place made me grin so much my cheeks hurt.

Scientists formerly believed that Tikal was a ceremonial center where only priests and acolytes went. Now they believe that Tikal was actually an urban center for trade and commerce. Mayans practiced *slash-and-burn agriculture*, and corn was their principal crop. Classic Mayans practiced a variety of intensive agricultural techniques, including ridging, terracing, and raising fields in swampy areas. Slash-and-burn agriculture is an agricultural technique that has a limited life span in tropical climates. The initial cutting and burning of the forest releases nitrogen and other nutrients to the soil, but farmers can only get a limited number of crops out of shallow rain forest soils before nutrients run out. Scientists often wonder if the Mayans declined because they exhausted the soils that surrounded their classic centers.

Warm Words

Slash-and-burn agriculture is an agricultural practice in rain forest regions where the forest is cut down and burned so that nitrogen and other plant nutrients in the trees are released into the soils for crops.

But scientists examining sediment cores from Lake Chichancanab in the Yucatan believe that drought might have brought the Mayan civilization to its knees. The core represents a 2,600-year-long record of the world the Mayans had to contend with.

Climate may have felled the Mayans at the height of their culture. Evidence shows that drought peppered the 2,600-year record, but the most intense and most prolonged drought ran from about 750 to 900 C.E. The onset of this drought period would have resulted in crop failure, famine, and disease. The increased competition for food may have lead to revolts against the ruling classes, as well as war between neighbor states for limited resources.

Anasazi ruins.

(Photo by Michael Tennesen)

The Anasazi

The Anasazi Indians, whose elaborate structures flourished in the Southwestern United States, may be a similar case. In what is now New Mexico and Arizona, they built an elaborate complex of roads, irrigation channels, and giant pueblos of stone and masonry. They built 75 towns in Chaco Canyon in New Mexico alone. The great house of Pueblo Bonito, perhaps the largest of the Chaco pueblos, appeared around 1000 C.E. Supported by massive timbers, its adobe walls rose four stories high and encompassed 800 rooms. Around these structures were a number of circular excavations known as *kivas* that were used as ceremonial chambers.

At Mesa Verde in Colorado, they built homes into the sides of cliffs. The largest dwelling at Mesa Verde, called "Cliff Palace," is three stories high with 200 rooms and several towers.

Warm Words

Kivas were the ceremonial chambers that surrounded the towns and cliff dwellings of the Anasazi Indians. Their structure was influenced by an earlier day when these Native Americans built homes of covered pits.

It probably housed 250 people, each family living in a one-room house. The Indians climbed up to their cliff houses using finger and toe holes carved into the sides of the cliffs. The difficult ascent was a strategy to keep their enemies out.

But then in the late 1200s, the Anasazi abandoned these two sites. Tree ring evidence gathered from the massive beams that were used to build some of the later structures again point to drought as the principal culprit in this enduring mystery. In the late 1200s little rain fell in the region, and the Anasazi might have abandoned their great cities to flee south to areas around the Rio Grande River where water was more reliable. Soil erosion, warfare, and disease might have also played a part. All this was a result of climate change.

Vikings Don't Like Cold

Around 800 C.E. the Vikings ventured out onto the Atlantic with their new vessels to pillage and explore. Their mastery of sailing gave them an edge over the Europeans. A warming trend occurred from about 950 to 1250 C.E. in Europe, the north Atlantic, and North America. Records from the Greenland ice cores and the California tree rings confirm this. The Vikings soon spread north into the Arctic, west to Greenland, and ultimately to Newfoundland. They did this in their early days without so much as a compass. But it was a relatively warm period, one free from severe storms, and accompanied by a long retreat of the sea ice. At its peak the Greenland settlement had over 280 farms and a population of around 3,000.

Climatoids

A book written sround 1125 describes the Viking settlement in Greenland toward the end of the first millennium. Then a man named Thorkel Farserk, a cousin of Erik the Red who founded the Greenland Colony, who had no boat at hand, swam out to an offshore island to pick up a full-grown sheep and carry it home to entertain his cousin. English scientists believe that 50°F (10°C) would be about the lowest even a fat person, not trained in long distance swimming, could swim the distance mentioned. The average temperature in modern times for that water seldom exceeds 43°F (6°C) so it seems the water must have been at least 7°F (4°C) warmer during the time Farserk swam for his sheep.

The warmth did not last. The Viking settlement gradually contracted as the climate cooled into the Little Ice Age that followed. The settlers brought their farm animals into their homes when the winters got too cold. Eventually they ate those farm animals as Greenland was increasingly locked in ice. In the late 1400s Pope Alexander VI noted that no ship had put into shore on Greenland in decades and the colony was almost unreachable. The Vikings eventually died out, though their Eskimo neighbors, the Thule Inuit, survived the cold just fine.

The Little Ice Age

From about 1400 to 1850 C.E. much of Europe, North America, and other parts of the world experienced what scientists refer to as the Little Ice Age. According to Richard Alley, it was not a stupendous plunge back into the ice age as was the Younger Dryas event, but it was a taste of the instability that is inherent in the earth's climate. Many regions experienced harsher winters than they'd known before.

Beginning in the early 1600s, the Thames River in London began to freeze occasionally. *Frost fairs* were held during winters when the ice was solid enough to support festivities. The fair included a small tent city replete with food, festivities, and games like ice bowling and ice skating.

The cold winters affected French vineyards, held up traffic on the Dutch canals, and changed the course of wars. It created harsh conditions for the American troops at Valley Forge. It might have been the Achilles' heel for Napoleon on his ill-fated march into Russia.

Warm Words

From the beginning of the 1600s to the beginning of the 1800s, during a time when much of the world was enduring the Little Ice Age, **frost fairs** were held on the Thames River when the river froze over. Since then, the world has gotten warmer, and now the Thames seldom freezes.

Storms were frequent and nasty. The North Sea floods produced a tremendous loss of life in 1570. A storm in 1634 permanently changed the coasts of Denmark, Germany, and the Netherlands. The great storm of southern England in 1703 blew down the lighthouse near Plymouth, uprooted enormous trees, and demolished towns and the country all across England. Ships were blown upriver or wrecked on the coast or at sea. London sustained millions of dollars worth of damage. The storm took more than 8,000 lives.

The biggest difficulty in the Little Ice Age was not so much the cold, but the storminess and erratic nature of the colder climate. Deep and sudden chills destroyed harvests. In the summer of 1665 and 1666, in the middle of the worst century of the chill, the weather turned extremely hot. London experienced its last great epidemic of the plague. At the end of the summer of 1666, a great fire burned through the city. During the Little Ice Age, temperatures averaging 3.6°F (2°C) less than in modern times shortened the growing season in England five weeks on average. Though 3.6°F

(2°C) may not sound like much, it points to what a small shift in the average temperature can do. (Compare these figures to the 2.5 to 10.4°F (1.4 to 5.8°C) of warming scientists predict for the next century.)

In North America, eighteenth-century winters impacted the lives of New Englanders far more than winters do today. When the waters froze, boats could not fish or haul goods. Deer that were used for meat died in heavy snowstorms, in which snow piled up and made it difficult for the deer to escape wolves or bears. In 1717, four snowstorms blanketed the region from February 27 to March 7 with five feet of snow. Winds then whipped the snow into drifts ten to twenty-five feet high.

In the winter of 1740–1741, Boston Harbor became such an expanse of thick ice that sleighs carried Sunday worshipers across it from late December through early April. One man then traveled 200 miles over the ice from Cape Cod to New York City.

The Dust Bowl

The climate started to warm up in the 1850s. By the second and third decades of the twentieth century, the climatic warming became evident to everyone. One of the tragedies of the warmer weather was the 1930s phenomenon known as the *dust bowl*. The dust bowl was an agricultural region that spread over 150,000 square miles (390,000 square kilometers) of the plains states of Kansas, Colorado, New Mexico, Texas, and Oklahoma.

Warm Words

The **dust bowl** was an area in the south central plains of the United States that was devastated by drought in the 1930s.

Warm Words

Black blizzards is a term that was used to describe the dust storms that swept the south central plains of the United States during the dust bowl era of the 1930s.

Cattle and other stock had been the principal business in this area until World War I. Then grain prices jumped and farmers plowed up millions of acres of grassland and planted wheat. But then the Depression hit, followed by a devastating drought from 1933–1939. With the demise of the grasslands, cultivated fields—harvested of their wheat and dry from years without rain—were an easy target for the winds that ripped across the plains. Wind-blown sands piled up in dunes, some 30 feet (9 meters) high. They blocked roads and choked pasture grasses. Lighter soils were blown into dust clouds that were 5 miles (8 kilometers) high. They formed *black blizzards* that swept all the way to the Atlantic Coast.

Wind erosion on this scale was something farmers in the area hadn't experienced before. Many pulled up stakes and moved to California. John Steinbeck catalogued some of the misery of this era in his novel *The Grapes of Wrath*, when thousands of people migrated from the south central states to California, desperately searching for work. Others stayed to salvage the land. Federal and local soil conservation districts taught farmers how to rehabilitate the soil, and millions of acres were restored to grass. But farmers' memories were apparently short-term, and when rains returned in the 1940s, many plowed up the grasslands again. In the 1950s, when another drought hit, Congress moved in and paid farmers to restore wheat land to grass and to instill soil conservation measures. These included alternating fields each year to restore moisture, using drought-resistant cover crops, and planting rows of trees to check the wind.

Changes Present

The twentieth century began hot, cooled a little in the middle (possibly from all our own pollution), and is rising rapidly once again. Many scientists tell us it is caused by carbon dioxide and other manmade greenhouse gases in the atmosphere. So what has the modern warm period brought us so far?

The National Oceanic and Atmospheric Administration reports a reduced diurnal temperature range, the difference between daily high and low temperatures, over about 50 percent of the continents since the middle of the twentieth century.

Who's Got the Ice?

The past century has seen a dramatic retreat of glaciers worldwide. In an engraving made in the mid-1800s of the French village of Argentiere, a glacier looms large over the town. In a photo in 1966, the river of ice has pulled well back into the mountain.

The use of nuclear submarines by scientists since the close of the cold war has allowed us to explore the Arctic Ocean in ways we never could before. The SCICEX (Science Ice Expeditions) programs spearheaded by the U.S. Navy and the National Science Foundation has launched a number of undersea expeditions in recent years below the giant Arctic ice cap. They report that the Arctic Ocean's ice cover has thinned by an average of four feet—about 40 percent—since the 1960s!

If the Arctic continues to warm, the consequences could be grave. It is possible that the ocean's summer ice cover could completely melt in the coming decades. Although the possibility is remote, the consequences could be catastrophic. Climate models show that the absence of ice in the Arctic could completely change climate patterns in the northern hemisphere.

Changes in El Niño

And there are problems brewing down south. In 1976–1979 the behavior of El Niño in the south Pacific off the coast of South America changed its tune. (For more information see Chapter 11.) El Niños suddenly got more frequent and intense, and the cooler La Niña patterns dissipated. It is unclear if this is due to global warming, but scientists believe that if global warming continues, El Niños will get even worse.

The Least You Need to Know

- ◆ Ice age hunters only converted to agriculture after things got warmer.

- ◆ Though the weather has been nice for 10,000 years, small changes have made big differences.

- ◆ Even Julius Caesar had to wait till the winds died down to cross the English Channel.

- ◆ Weather promoted the plague.

- ◆ The Little Ice Age (1400–1850) was a time of chill.

- ◆ In the twentieth century, the ice is melting and El Niño is getting crazier.

Chapter 8

The Dawn of the Industrial Age

In This Chapter

- ◆ Britain stokes the Industrial Revolution
- ◆ The steam engine hauls the loads
- ◆ Back in the USSR
- ◆ Riding in cars
- ◆ A Grand Canyon haze

The Industrial Revolution has brought most of the manmade changes in our atmosphere. The term was first applied to the period in Britain from about 1750 to 1850. During those years the steam engine and the rise of industrialization radically transformed commerce and society. The new machines could do the work of many hands, and economic output grew at phenomenal rates. From Britain, the revolution spread to Europe, North America, and is under way right now in various nations in the southern hemisphere. The creation of metal, machines, and tools has fueled changes in society and in the atmosphere that are profound. I'll discuss them in this chapter.

Early Changes in Britain

The Industrial Revolution had its roots in the scientific progress of the *Renaissance* (from the 1300s to the 1500s). Leonardo da Vinci sketched the precursors to the machines that would later be invented during the Industrial Revolution. Factories in Sweden were using waterpower as early as the 1720s. Gunsmiths in France were then developing their own factory system. But it was in Britain that all the changes came together in the middle of the 1700s.

Feudalism began to break down after the Renaissance. The self-sufficient manorial village with its lords and peasants gradually gave way to commercial farming where farmers took their products to markets to trade for cash. Market considerations replaced traditional practices and things got more efficient.

Warm Words

The **Renaissance** refers to the great revival of art, literature, and learning that occurred in Europe from the 1300s through the 1500s.

In medieval times when repeated plantings exhausted the soils, farmers would let as much as one third of their land lie fallow so that the soils could recover. However, in the 1700s British farmers began to use crop rotation, planting beans and grasses which put nitrogen back into the soil. Property that previously lay empty part of the time could now be used continuously.

Crop rotation allowed farmers to use the land that had previously lain fallow for pasture since the cattle could graze on the grasses. Thus farmers didn't have to dedicate land strictly for pasture and could plow up old pastures, grow more crops, and raise more cattle. With more cattle they had more meat to get through the winter, more leather, and more organic fertilizer.

Then they started putting up fences. English agriculture had previously run under an open field system, in which everyone in the village had equal rights to the open land. But Parliament let people stick up fences around their lands, so things got more competitive and, again, more efficient. The agricultural revolution set the stage for the industrial one to follow.

The rise of free market trade happened in Britain much more rapidly than it did on the continent. The British nobility was hooked on trade. Britain had lots of rivers and seaports and an overseas empire with plenty of raw materials. Merchants, landowners, and ship captains were out hustling, investing, and doing deals. Improvements in sanitation, health care, and agriculture helped the population expand. Europe had money to burn from the gold and silver it had stolen from the newly conquered Native Americans. British merchants looked at each other and said, we've got to have ourselves a sale.

Climatoids

Why Britain? The Industrial Revolution got started first in Britain for a number of reasons: The British nobility was interested in trading goods, and the country built lots of waterways to move the goods around. The British textile industry was hungry for ways to mechanize its processes. James Watt, a Scottish engineer, discovered the steam engine, and one of its first uses was to stoke the fires to produce more iron. They used the iron to make machines. While the continent was resisting change, Britain was forging ahead.

The Industrial Revolution has changed our atmosphere.

(Photo by Michael Tennesen)

Textiles Lead the Way

To have a sale, however, the British had to have something to sell. Textiles must have certainly come to mind. The British were famous for it. But making cloth and clothing was such a slow process. There are a number of tedious steps that go into producing a piece of cloth. The fibers must be combed until they are parallel. They need to be spun or twisted to make yarn or thread. After which you had to get out your loom and hand-make some cloth.

People had been trying for centuries to mechanize these processes. Lewis Paul invented a spinning machine in 1738 that could mechanically shape fibers and spin thread. Edmund Cartwright patented a crude power loom in 1785. The handloom operators burned down one of the first factories, but that didn't stop progress. By 1813, there were about 3,000 power looms in operation in Britain; 20 years later there were 100,000.

The Steam Engine

It was the steam engine that really fueled the Industrial Revolution. By the early 1700s there were already a number of crude steam-powered machines for pumping water out of the mines. (Mines flood when they go deeper than the water table.) But these early devices used massive amounts of coal and were only practical at mine pits where coal was cheap. James Watt (1736–1819), a Scottish engineer, was called to fix one of these big guys in the 1760s and he started thinking about ways to make it better. With the old engine, the cylinder had to be heated up and then cooled down to bring about condensation. Watt devised a separate condenser attached to the cylinder. He made the engine reciprocating, by letting steam into one end of the cylinder and then into the other end, adapting this power to produce rotary motion. In other words, it could turn a wheel.

More than that, by converting the steam power into rotary power it could turn mills and operate machinery that had previously been powered by water wheels. Plus, the machine could be taken anywhere you wanted and you didn't have to rely on the local creek to keep going.

Coal and Iron

Before 1800, most machinery was made of wood. But wood splintered, warped, and cracked. If the Industrial Revolution was to get going it needed some harder material for its equipment.

There was iron, but iron wasn't easy to create on a mass scale. Iron ore contains iron oxide. You have to get the oxygen out to make metal out of it. Up until then, iron was refined with charcoal fires. As the charcoal burned beneath the ore, it drew the oxygen off and left the pure metal. But charcoal was made from wood, and the forests of Britain were going down fast. So it was a big deal in the eighteenth century when the iron makers figured out how to use coal to make their iron.

With coal you could get your iron molten but it still had some impurities, chiefly carbon. This was wrought iron, and it could be poured into molds, but it was too brittle to make into hinges, swords, and hardware. To get the better grade of iron necessary for tools you had to get the fire hotter, usually by blowing air on it to produce carbon-free metal. One of the first uses for the steam engine was to power these blowers to improve the production of iron. Steam engines were also used to power forging hammers and rolling mills. It wasn't until the 1850s that industrialists learned how to convert iron metal into the harder steel.

Climatoids

Though the Industrial Revolution replaced workers on the line, it created a whole new group of craftsmen in the toolmakers. These men created the machinery, tools, and instruments used to get the revolution going. With a new set of tools in the boring mill, engine lathe, and planing machine, they created such finely crafted metal parts and machine products that today their handicrafts are considered by many to be works of art. Some of the finest workmanship was applied to the telescopes that appeared in the 1800s. Modern astronomy owes its existence to the telescope maker and instrument makers of the time.

What's That Smell?

It wasn't long before the new factories began affecting the air that people breathed. But it was not the first time there had been pollution problems in Britain. As far back as the thirteenth century, Londoners had complained about the tanning, pottery, and lime production industries, all of which contributed their particular smells to the city air. Coal came to London soon after 1200 C.E., and mounting sewage problems made their contribution as well. In 1257 King Henry III's queen became the champion of the complainers—one of the first environmentalists—and from that time on commissions were regularly established to consider air pollution.

However, as the forest dwindled and wood became more difficult to find for domestic fires, people turned more to coal. Nice people at first refused to enter rooms where coal had been burnt, but the snobbery died quickly as more people started using the black substance.

During the 1600s, rickets, a children's disease caused by a lack of sunlight, became increasingly common due to smoke-filled skies and factory labor for children. The 1700s and the 1800s didn't get any better. A survey conducted in the early 1900s found that half the children in the poorest districts had rickets. Other major European cities experienced similar problems when the Industrial Revolution was exported to their countries. But the increased capital that industrialization brought with it was apparently too enticing for people to object to a lack of sunlight. Rather than complain about pollution, they simply learned to live with it.

Global Warnings

You can chart the degradation of the air during the Industrial Revolution by looking at the art it created. Blue skies dominated the backgrounds of most of the early landscape painters in Britain and the continent. But as the decades progressed, backgrounds became hazier and the blue disappeared.

Steam Transportation

The Industrial Revolution really started moving when the steam engine was put on a railroad car. Rails laid on prepared beds had been used before, principally for the horse-drawn cars used for hauling ore and heavy machinery around the mines. It wasn't until the early 1800s that the first steam-powered locomotives were in use. It was a revolution in shipping. No longer did industry have to depend on the pack-horse or the wagon. In 1807, Robert Fulton, an American who'd been educated in Britain, built the first steamboat and launched it on the Hudson River.

Britain wasn't afraid to borrow ingenuity from its former colony. There followed a heavy spurt of canal building in the British Isles. Canals linked many parts of the country and moved heavy freight at a fraction of the cost of hauling it overland. Roads kept improving as well, particularly after the first iron bridge was developed in the late 1700s.

Industrialization Spreads

In the nineteenth century, Britain was still the center of the mechanized miracle. At first it only exported products of the Industrial Revolution. It was the proverbial merchant with lots of goods to offer, but little information on how it developed those products. However, it was a secret too lucrative to be kept. Soon the revolution spread to Europe and North America. Then it spread to Japan, China, and India.

Pollution followed the spread. Visible smoke and the noxious fumes of sulfur dioxide (SO_2), a byproduct of the burning of coal, followed wherever industrialization went. London fogs, a mixture of fog and pollution, were at times so thick that they required street lamps be turned on in the middle of the day. Dickens called these fogs "London Particular." As pollution spread, so did the resistance to it. Laws on air quality were adopted as early as 1815 in Pittsburgh, Chicago, and Cincinnati.

By the nineteenth century the citizens of London had banded together to protest the increasing problem. But the cries of those complaining were drowned out by those who'd tasted the fruits of industrialization and wanted more. And the same story repeated itself in Europe and the United States. Industrialization moved forward.

The Car

A French engineer built the first motorized vehicle in the late 1700s. It was a heavy steam-driven carriage with a big boiler that stuck out in front. It could travel a whopping 3 mph (5 kph). The Stanley brothers of Massachusetts, the best-known

American manufacturers of steam-driven cars, produced their famous Stanley Steamers from 1897 until after World War I.

The rise of the automobile was pushed forward by the development of the internal combustion engine. Karl Benz, a German engineer, built the first vehicle of this type in 1885. French, German, and U.S imitations followed. In the United States, the first horseless buggies were manufactured in the 1890s. Many of the early engines had only one cylinder, and a chain and sprocket drive that pushed wooden carriage wheels. They were open, like the buggy, accommodated two passengers, and were steered with a lever.

Automobile production didn't take off right away. George Selden held a patent on the automobile with a group of manufacturers. Henry Ford, however, represented a group of independent manufacturers who refused to acknowledge the patent. Selden's group sued Ford and his cohorts. In 1911 the U.S. Circuit Court of Appeals ruled that the patent only covered the two-cycle engine. Ford's engine was a four-cycle engine. Ford then got the assembly line going and soon the automobile was linking cities and suburbs all across North America. Trucks brought luxuries to places that hadn't known them before. Together cars and trucks began to transform the landscape and society, as well as the air.

The Ruhr and World War

Industrialization heated up on the continent, particularly around both World Wars. Much of the activity in Germany was centered in the Ruhr region, near the junction of the Ruhr and Rhine Rivers. In 1850 this had been an agricultural area, but then a rich seam of coal was discovered beneath it. By 1910, the region employed hundreds of thousands of miners, digging up millions of tons of high-sulfur coal. The coal sustained the giant steel- and ironworks of Krupp and Thysen, which were the muscles behind the German military-industrial complex.

At the start of the twentieth century, the Ruhr was the biggest industrial region in Europe. Because this industry was so vital to the German military machine, it escaped most regulation. Soon the air was polluted with smoke, soot, and sulfur dioxide.

Under Hitler, Ruhr factories stoked their furnaces even higher. Though the Nazis expressed affection for German blood, that affection apparently didn't extend to German lungs or German air. In 1944 and 1945, Allied bombers made special targets of the Ruhr, eventually leveling a large part of its factories. The haze, however, made Allied bombers far less accurate.

Global Warnings _____

Throughout much of the first half of the twentieth century, the Ruhr region, the industrial heartland of the German military machine, was so polluted that many came to believe that pollution was natural. In 1923, French and German troops occupied the Ruhr in response to Germany's failure to pay war reparations. Strikes followed and the industry was shut down over the summer. The sky got bluer, trees grew faster, and the local harvest improved by half. There really was life without pollution. But then strikers settled, went back to work, and everything got hazy again. An official inquiry into the polluted air, inspired by the blue summer of 1923, concluded that the pollution was inevitable, and warned that the people must adapt, rather than try and control industry.

Communism and Pollution

In the 1960s, air pollution became an important political issue in both Germany and in North America, but not in the USSR or any of its eastern European satellites. High billowing smokestacks were a symbol of pride for Marxist regimes—a promise of power and growth for its society. Iron, steel, and coal were the cornerstones of the economic plans of the Communist countries.

Air pollution was the norm. It was viewed as a symbol of progress rather than a plague of industry. Citizen pressure and pollution control, which were growing in the Ruhr toward the later part of the twentieth century, were absent behind the Iron Curtain. The state protected its polluters. Popular dissent wasn't allowed behind the Iron Curtain. It wasn't until the 1990s, after the fall of communism in eastern Europe and Russia, that some of the cries against pollution were finally heard.

Hot Debates

In the 1960s when the German citizenry of the Ruhr Valley began to complain about pollution, the first response by industry was to put up colossal smokestacks that had the immediate result of reducing the emissions locally, but broadcasting them over a much larger area. During the Communist rule of the USSR and eastern Europe, these tall smokestacks were perceived by the Soviets as a symbol of power. Their usefulness, however, was debatable. In the end, they might have simply helped to spread the pollution problem to a wider audience.

The Japanese Story

Japan followed the United States and Britain and from the 1870s on built heavy industrial centers. The nationalism of Japan and Germany in the early part of the twentieth century promoted industrialization as it expected the citizenry to endure pollution for the good of the national cause. Both countries quickly became some of the most polluted in the world. However, some complaints were heard. Local farmers got miners to curtail operations during the rice harvest. In response to local complaints, the Hitachi copper mine in the 1910s built the tallest smokestack (490 feet, 155 meters) in the world.

The automobile took over in the second half of the twentieth century. Tokyo's car fleet quadrupled from 1960 to 1970. When the head of the American EPA went before Congress to push the Clean Air Act in 1970, he showed them photos of Tokyo traffic cops who had to wear masks to filter the air. It was an example of how bad things could get in the United States if Congress didn't enact tougher regulations.

But by 1990, Japan was not the same polluted nightmare it had once been. Part of the reduction in pollution came from the switch from coal to oil and other energy sources. The Japanese economy was then booming and Japanese industry had enough capital to invest in pollution control. Industry focused on energy efficiency and began to use less fuel even as it was growing.

When the U.S. Clean Air Act was being enforced in the 1970s, Japan's powerful Ministry of International Trade and Industry came up with its own environmental restrictions. But whereas American auto-makers sued the EPA, Japanese auto-makers complained but then dived into the task of meeting both the new Japanese and American standards. The result was that by 1978, new Japanese cars emitted only about 10 percent of the pollution of the 1968 models. While the major American auto-makers dallied, Japanese car manufacturers surged into the American market in the 1970s.

CAUTION

Global Warnings

Estimates vary as to the effect air pollution has had on the health of the world's population during the last century. In 1996, the Harvard School of Public Health estimated that air pollution killed 568,000 people annually. In 1997, the World Health Organization estimated that the annual world death toll was closer to 400,000. One estimate of the human toll of pollution during the twentieth century puts the total deaths at between 25 to 40 million, roughly equal to the combined casualties of World Wars I and II.

Acid Rain

Acid rain, derived from sulfur and nitrogen oxides, became an international issue with neighboring countries in the twentieth century. Unlike most pollutants, sulfur and nitrogen oxides are long-lasting in the air and are capable of traveling thousands of miles. The combustion of fossil fuels in the industrial centers of the twentieth century generated huge sulfur emissions, the main ingredient in acid rain. Sulfur and nitrogen oxides released into the atmosphere by the burning of fossil fuels combines with water vapor to produce sulfuric acid and nitric acid. Acid rain comprised of these elements can drift out over larger areas and decimate forests, fish, and fresh waters.

Warm Words

Acid rain is a general term for many phenomena including acid rain, acid fog, acid sleet, and acid snow. It is produced when emissions of sulfur and nitrogen from the burning of fossil fuels combine with the moisture in the air to produce a mild solution of sulfuric acid and nitric acid that is destructive to plant life, aquatic life, and buildings.

One of the consequences of industrialization is the rise of the unhappy downwind club whereby nations who don't reap any of the benefits of their industrial neighbors nevertheless share in their pollution—principally delivered as acid rain. In the 1990s, industrial pollution from Korea and China rained down on Japan. Acid rain produced around the Great Lakes and in the Ohio River Valley fell on Quebec (as well as New York and New England). It was the same case in the 1960s, when British pollution spoiled the rivers and lakes of Scandinavia. In 1985 and 1994, most European countries signed protocols requiring sulfur emission reductions. Emissions have dropped 15 percent from 1980 to 1995, driven mainly by the energy efficiency stimulated by petroleum price hikes in the 1970s.

Global Warnings

Acid rain and the acidic brew that comes from smokestacks and tailpipes corrode limestone and marble and have thus taken a toll on our cultural monuments during the twentieth century. Acidic London fogs have eaten away at the stone facing of St. Paul's Cathedral. Athenian pollution has done more damage to the ancient marble on the Acropolis in the last 25 years than all the weathering of the previous 2,400 years. Michelangelo's David could no longer be trusted to Florence's acid air and was replaced by a replica outdoors. And the Sphinx, which has survived centuries of desert sandblasting, has shown its greatest degradation in the twentieth century as a result of Cairo's pollution. How many of these monuments will even be recognizable in the decades to come?

Smog

The experience of Los Angeles in the 1940s was a harbinger of pollution problems that would occur elsewhere in big cities across the nation. When the residents of Los

Angeles complained of the smog, few people were aware at that time that smog—which derived its original meaning from a combination of smoke and fog in London—was indeed different there. Los Angeles industries did not use coal, yet the air was still hazy with pollution. What scientists discovered was that Los Angeles smog wasn't due to smoke and fog, but to the action of sunlight on gases emitted chiefly from car exhausts.

Los Angeles had gotten rid of its trolley system in the early part of the twentieth century and put its attention on a system of freeways that emphasized the car as the chief mode of transportation. As air pollution problems have increased, the city has tried to reinstall its trains as well as add subways and buses. But the expenses are far greater since the city tore up its rails and in many cases sold the right of ways. Smog is also a problem in the Baltimore to Boston urban corridor, but at least on the East Coast, mass transit is well established and need not be reborn.

Our National Parks

From the Great Smoky Mountains to the Grand Canyon, air pollution is seriously affecting our national parks. Clouds over the Smoky Mountains can be 1,000 times more acidic than natural rainfall. And more than 30 species of plants in the park show injury from ozone. At the Grand Canyon in Arizona, pollution spoils the views on 90 percent of the days. Visitors to the south rim can only see half as far as when the park was dedicated in 1919.

Figuring out where it all comes from is not an easy task. At the Great Smoky Mountains, pollution can originate from power plants in the Tennessee or the Ohio River Valleys, or it can drift in at high altitudes all the way from Atlanta or New York. A recent study in the Grand Canyon showed its pollution came from industry in Arizona, cars in California, and smoke stacks in New Mexico, Nevada, Utah, and unregulated pollution sources as far away as Mexico.

Et Tu Global Warming

And now we add global warming to the mix. In many cases it's the same band of desperados that are the villains of ground-level pollution: ozone, carbon monoxide, and particulates, as well as sulfur oxides. All have greenhouse effects. But the principle villain of the global warming problem is carbon dioxide. Unlike the other pollutants, it is colorless and odorless, but is far longer lasting.

Pollution can change the weather locally and also in areas that are downwind from pollution sources. Suspended particulates act as condensation nuclei upon which the moisture in the air gathers to form clouds, rain, snow, hail, and fog. In Pittsburgh and

Buffalo, rainfall is generally lower on Sundays when industry and automobiles are running at lower volumes as well. In Toronto, Canada, the snow is heavier when there is more particulate matter in the sky.

However, as bad as our pollution problems are, they do not have the capacity to change the climate, as does global warming. Yet pollution comes from the same source as global warming. It is the result of the same growth in industry, the same automation, and the same cars that produce the pollution. It is another bastard child delivered up by our use of fossil fuels.

The Least You Need to Know

♦ The Industrial Revolution occurred first in Britain from 1750 to 1850 and has since spread throughout the world.

♦ Both the Nazi and the Communist regimes encouraged industry at the sacrifice of the air.

♦ Japan experienced a reversal of pollution problems in the latter half of the twentieth century.

♦ Pollution has no problem jumping international borders.

♦ Pollution and the manmade greenhouse gases responsible for global warming have the same source—the burning of fossil fuels.

The Axe and the Plow

In This Chapter

- ◆ Forests covered the land
- ◆ Man removes the covers
- ◆ Rainforest wealth
- ◆ Rainforest devaluation

Man has been hacking back the forest since he first decided to take up farming and ranching. There are various estimates of the history of forest destruction by man given by environmental groups. Anywhere from 20 to 80 percent of the forest that existed in preagricultural times has been removed. Today the biggest battles between man and the trees are taking place in the tropical rainforest. Rainforests are found around the world, mostly near the equator. About 80 to 90 acres of tropical rainforest are destroyed every minute. An area the size of New York disappears each year. And as the trees are cut down, they release a tremendous amount of carbon dioxide into the atmosphere. The destruction of the forest and the tilling of the land produces about 25 percent of the greenhouse gasses released by man into the atmosphere.

The Rise of the Forest

Forests have been on the earth millions of years, although their ancestors looked quite different. The fossil remains of the scale tree, actually a form of giant moss, are found in the rock strata of the Carboniferous period about 300 million years ago. These guys had weirdly shaped skinny stems that could reach 100 feet (30 meters) in height. They occurred in the warm moist lowlands that surrounded inland seas.

Warm Words

Conifers are trees with needle-shaped leaves that are evergreen—they stay green all year. **Broadleaf** trees are those with broad, flat leaves that are often deciduous—they shed their leaves each year.

The first time that trees start to look anything like what we see today was about 100 million years later. *Conifers* and *broadleaf* trees began to form and diversify. Conifers are evergreen trees (they stay green all year) and have needle-shaped leaves. Broadleaf forests are generally deciduous trees (they shed their leaves each year) and have broad, flat leaves. Conifer and broadleaf trees represent the two major categories of trees in the world. Conifers are found in cooler, drier climates, and broadleaf forests are found in warmer, wetter areas of the world.

Original Cover

During the last ice age, the forests covered only those areas not under ice. But as the glaciers retreated, the forests advanced toward the poles and up the mountains. Scientists believe that the world was moister coming out of the last ice age than it is today. Pine, oak, maple, olive, and other species forested northern Africa. By about 2000 B.C.E. this Mediterranean type of forest cover was gradually replaced by open semi-desert vegetation. In North America the coniferous forest advanced northward after the last ice age, chased from behind by temperate broadleaf species.

Forest Types

About a third of the forested area of the world is coniferous forest, most of it occurring in the northern hemisphere. The area that encompasses the former Soviet Union contains 45 percent of the world's coniferous forest, while North America contains about 35 percent. The continents of South America and Africa alone contain about three fourths of the world's broadleaf forest, much of it being tropical rainforest.

The forests have evolved a number of unique ways to spread their seeds. Species with light seeds like willows can be dispersed by the wind or carried downstream by water. Many trees grow fruits, which they use as a bribe to get others to disperse their seeds for them. Animals and birds eat the fruits, which contain the seeds, and then disperse the seeds in droppings. In this way, bats disperse almost 50 percent of the trees in some tropical forests. Conifers release their seeds in cones, dropping them to the forest floor. Some species of pine are programmed by nature to release the seeds inside their cones only when heated by fire. This gives them a jump start on the competition following a forest fire.

Carbon Dioxide and the Forest

Carbon is one of the most important elements of the forest. Trees and other plants absorb carbon dioxide, convert it to food and energy, and release oxygen back into the atmosphere, so all of us oxygen-consuming creatures can keep sucking up air. We take in oxygen and exhale carbon dioxide. This cycle makes sure that the supply of carbon dioxide and oxygen remains fairly stable. But cutting down vast numbers of trees, whether in temperate or tropical forests, releases billions of tons of carbon dioxide into the atmosphere.

Forests act as carbon sinks, or storehouses of carbon, removing vast quantities of carbon dioxide from the atmosphere. One of the ideas proposed in the 1997 Kyoto Protocol was to allow industrialized nations to meet their carbon dioxide reduction goals by planting more trees. Instead of having to reduce fuel emissions by 10 percent, they could reduce emissions by 5 percent and plant enough tress to take up the difference. (For more details refer to Chapter 19.)

Old World Destruction

With the transition from nomadic hunting and gathering to agriculture, man began cutting down the forest to make room for crops. As civilization developed, so did the need for timber. Man used wood for fuel, shelter, and other domestic uses. Wood was important to building mines and smelting metal. It was used early to build dugout canoes and later for tall ships.

At the end of the first millennium, settlers had made deep inroads into European forests. Wood in the Middle Ages was an important part of the British military machine. Each foot soldier carried 24 arrows made from the wood of ash trees harvested specifically for that purpose. Bows were made of yew. By the 1600s, French officials were trying to regulate wood harvests so that enough good quality wood was left over for shipbuilding. During the 1600s and the 1700s, woodsmen were already looking for ways to reproduce and sustain the forests.

Trees got in the way.

(Photo by Michael Tennesen)

Settling America

But when the settlers reached America, their first thought was clearing the land for farms, not preserving the trees. Native Americans had a much better sustainable relationship with the forest and the land. But the new settlers had farms to clear, grazing lands to develop, and industrial fires to stoke.

Clearing the Land

There were approximately 850 million acres of forest when European settlers first arrived in America, but quickly about 100 million acres were cut or burned to make way for farming. Between 1850 and 1900, while the United States was engaged in its own Industrial Revolution, an average of 13.5 square miles was cleared every day to make room for farms. By 1900, agriculture, construction, and fuel needs had reduced the original forest cover to less than 580 million acres.

Prior to 1900, North Americans were not that interested in preserving forest. Their main interests were expanding agriculture, roads, industry, transportation, and ranching. And all of these activities contributed to the destruction of the forests.

> **Hot Debates**
>
> The development of paper in the 1850s greatly expanded communication and the spreading of ideas in American and European societies. But the big drawback was the need for pulpwood to make the paper. Paper gave society another reason to cut down more trees.

The Lumber Industry

As America expanded westward, forest products became big business. Frederick Weyerhaeuser, who was known as the "Timber King," was born in Germany and moved to the United States in the 1850s. He married and then moved to Illinois, where he worked in a sawmill. By 1857, he'd saved enough money to buy the business. Things went so well that he bought several more sawmills along the Mississippi. In the 1890s he moved to Minnesota where he struck up a friendship with James Hill, the railroad tycoon. Hill had lots of forest that he'd bought cheaply from the U.S. government, principally as right-of-ways for his railroads, and he didn't know what to do with it. Weyerhaeuser talked Hill into selling him three million acres. The purchase seemed to wet Weyerhaeuser's appetite for timberland, a habit that eventually led him to own more forest than any other American. Weyerhaeuser's practices have come under fire by environmentalists in the Pacific Northwest and more recently in the tropical rainforest where the corporation has expanded its timber empire.

About the same time that Weyerhaeuser was gearing up his timber holdings, the U.S. government was getting involved in forestry practices to try and guarantee a sustainable yield of timber. Reforestation was first practiced in Germany in the 1300s. By the mid-1700s, it was well established in France, Switzerland, and Austria. German immigrants brought modern forestry practices to North America in the 1800s.

The timber industry has changed in recent years as large multinationals have taken over smaller timber companies in the Northwest and have sold off timber rapidly to pay debts or to finance other more profitable endeavors. Many timber companies now ship whole or split logs to Japan, Asia, and Mexico to be milled.

The Conservation Backlash

The idea of conserving our nation's forests got its first supporters toward the end of the 1800s. In 1892, the same year that John Muir established the Sierra Club, Congress added a rider to the U.S. General Revision Act, authorizing the president to set up forest reserves. Gifford Pinchot, one of America's leading advocates of forest conservation, was declared head of the Division of Forestry. Pinchot was born into a wealthy Connecticut family, graduated from Yale, and was encouraged by his father to get into forestry. Only no school in the United States or Canada taught forestry at that time, so Pinchot went off to France to see what he could learn.

In 1905, Teddy Roosevelt established the U.S. Forest Service and appointed Pinchot the chief forester. At that time, lumbermen regarded forest devastation as part of the job and the idea of rejuvenating the forest as a delusion. Pinchot emphasized long-term forest management and the wise use of timber resources. During Pinchot's five-year term as forest chief, the number of national forests more than doubled.

In the 1900s, excessive logging spawned controversy. Conservationists seemed particularly upset about *clear cutting*. Clear cutting is the practice of removing all the timber in a given area, rather than going in and selectively cutting trees. Environmentalists look on clear cuts as a blight. But clear cutting is necessary in areas where the type of forest needs lots of light for seedlings to regenerate. Some trees just don't reproduce well in the shade of others.

Warm Words

Clear cutting is the forestry practice of cleanly removing all trees in a given tract so light may rejuvenate seedlings. **Old growth forest** is a type of forest that has not been previously logged and represents a more natural and diversified mix of tree and plant species.

In the second half of the twentieth century, increasing attention to preserving *old growth forests*—forests that have never been logged—has created intense controversy between environmentalists and logging communities. Environmentalists argue that old growth trees are necessary to some species of animals, particularly the endangered spotted owl, which scientists believe need the more mixed, old growth forests to sustain themselves. Loggers argue that the real endangered species are the human logging communities that rely on cutting old growth forest for income. National parks and wilderness areas protect much of the old growth forests in North America today.

The Rainforest

Today the rainforests of the world are under the gravest assault. Tropical forests are vanishing at the rate of 250 acres per minute. This means that an area roughly equal to two football fields of rainforest (about 4 acres) disappears every second. An example is Mexico, which has lost half of its temperate and tropical forests since the 1960s.

Tropical rainforests are composed of three different types of forests, including tropical dry forest, tropical moist forest, and true tropical rainforest. True tropical rainforests receive about 80 to 120 inches (203 to 305 centimeters) of rain a year or more. Tropical rainforests cover about 6 to 7 percent of the earth's surface and are located in a band reaching 10 degrees in both directions from the equator and covering about 3.4 million square miles.

Tropical rainforests have layers of vegetation that form separate habitats for different animals. At the top of the forest is the upper canopy, a mass of treetops, vines, and flowering plants. This is where the eagles, hawks, sloths, and parrots hang out. The next level, the lower canopy, is home to birds, butterflies, and iguanas. The next layer is the understory, which begins above the forest floor. The forest floor is usually bare except for fallen leaves.

Wildlife Treasures

Tropical forests contain a larger diversity of plant and animal species than any other place on earth. Scientists estimate that anywhere from 50 to 90 percent of all animal species are found in tropical rainforests. Normally animals are spread out and hard to spot. They rarely congregate. I got to see an exception to this rule at Santa Rosa Park in Costa Rica.

I arrived there in the dry season, a period during which the park's several waterholes become increasingly important to its wildlife that gather there in numbers. In the early morning and late evening, I frequently walked a trail lined with black iguanas that led to one of these holes. Iguanas are giant lizards that can weigh up to 7 pounds (3 kilos).

One day as I approached the waterhole, a pungent odor filled the air. I looked to find a pack of peccaries (wild pigs) stoically drinking. Wild peccaries are said to be the most dangerous of Central American animals, but the peccaries at Santa Rosa were apparently nonaggressive. As I watched, a group of coatimundis, a South American cousin of the raccoon, threaded through the peccaries to share the water. Doves alighted on what little space was available while turkey vultures soared overhead. Then, as a chorus to this strange animal show, a troop of white face monkeys appeared, perching in the tree, waiting their turn for a drink.

If that watering hole hadn't been in a protected park, it might be farmland or a used and abandoned space by now. Tropical rainforest animals, many of which haven't yet been named by scientists, are disappearing as fast as their homes.

> **⚠ CAUTION Global Warnings**
>
> In 2001, for the first time since the 1700s, scientists declared a primate extinct. Primates are animals that are in the same taxonomic group as man. The primate was the Miss Waldron's red colobus, a West African monkey. It formerly lived in the rainforests of Ghana and the Ivory Coast. Extinction for nonprimates is now happening at a breakneck pace across the globe. Between 1 and 100 species become extinct every day. That's against a natural rate of extinction (without man's influence) of only 1 species per 100 years.

A Cornucopia of Pharmaceuticals

One of the little-known treasures of the tropical rainforest is its medicines. About 25 percent of all pharmaceuticals used by North Americans were originally derived from tropical plants. Quinine, used to treat malaria and pneumonia, was developed from the bark of a rainforest tree. Important ingredients in cortisone are produced from rainforest plants in Central America. Anti-cancer medicines, contraceptives, tumor inhibitors, anesthetics, and drugs to stimulate the heart and respiratory system have come from rainforest products.

The National Cancer Institute sends *ethnobotanists*, scientists that study the relationships between plants and human cultures, to rainforest countries to investigate the plants used by native healers to treat their patients. The institute is concerned that if the rainforest disappears, so may the healers *and* their medicinal secrets. Many of these cures have developed over thousands of years through practical application—try this, see if it works; if it doesn't, try something else. Though it's not the way western medicine likes to develop its pharmaceuticals, that doesn't mean that a number of valuable cures may not have developed from the process.

Warm Words

Ethnobotanists are scientists who study the cultural and biological relationships of native peoples and the plants in their environment.

In fact, there are currently more than 200 pharmaceutical companies in the world who are studying rainforest plants used by native healers and shamans to treat their communities. Many researchers believe the rainforest contains a treasure of chemicals that could be used as new pharmaceuticals.

Indigenous People

Indigenous people are among the rainforest's most precious riches. There are approximately 250 million indigenous people in 70 countries that control about 10 to 20 percent of the world's lands. Most of it is in the harshest regions: the Arctic tundra, the dry deserts, and the most virgin rainforest. About 50 million indigenous people live in the rainforests of the world. Contact with outsiders can have drastic consequences. Many die when Caucasian explorers enter their lands, carrying diseases for which the natives have no immunity. More than 30 percent of the Yanomami people in Venezuela have died from measles and whooping cough since their first contact with outsiders.

Many indigenous rainforest communities have a sustainable relationship with the forest that has lasted for thousands of years. According to the Rainforest Foundation of Great Britain, indigenous people practice shifting agriculture, hunting, gathering, and fishing. They have an intimate knowledge of plants that they use to provide them with food, medicine, pesticides, poisons, building materials, clothes, string, and implements.

Cutting It Down

Much of the fertile land in the Amazon Basin and the Philippines has been turned into large plantations to cultivate rice, sugar cane, bananas, pineapples, rubber, and coffee. The major timber companies have also gone into the rainforests and developed tree plantations which change the natural diversified forest into areas where single species are raised in an intense agricultural setting that includes chemical weeding, pesticide use, and mechanized harvesting. These processes take their toll on soil fertility and groundwater reserves, and increase erosion.

In the late 1960s, the Brazilian government began a program to encourage development in the Amazon that included building roads. They offered settlers 240-acre plots with guaranteed credits for planting crops, a six-month household subsidy, and food subsidies as insurance against disasters. To encourage commercial investment, they offered tax incentives to businesses, subsidized loans, and reduced import duties on agricultural equipment.

Global Warnings

The $40 billion Avanca Brazil (Advance Brazil) project announced by the Brazilian government in 2001 holds special dangers for the Amazon. The program promotes industrialization, highway construction, hydroelectric projects, and more. Researchers reported in *Science* magazine that the enormous project could degrade 28 to 42 percent of the Amazon rainforest by 2020.

Chico Mendez

Chico Mendez was born in 1944 in Brazil in an area of rainforest and ranchlands. His family worked on a rubber plantation and grew their own food or harvested it from the nearby forest. By the time he was in his early 20s, Mendez was organizing rubber tappers for better working conditions, fair prices from company stores, and the building of schools. When the government started building roads and cutting down the forest upon which Mendez depended, he joined with other grassroots leaders and several activist Catholic priests to organize the workers.

By the 1980s, the movement had grown so strong that a conference held in Brazil in 1984 attracted more than 4,000 delegates from unions and other organizations for workers. At the conference, Mendez promoted the idea of setting aside large sections of the rainforest as reserves. Mendes received the United Nation's Global 500 Award for environmental conservation in 1987. But his enemies were enraged at his successful efforts to stop rainforest destruction. So much so that in 1988 he was murdered by cattle ranchers. His death brought worldwide attention to the conservation of the rainforest.

Oil and Roads

In the last several decades, oil companies have explored various regions of the tropical rainforest in South America with the hopes of finding new reserves to feed the voracious petroleum appetites of North America and Europe. When oil companies find petroleum, not only do the derricks and pipelines pollute the soil and the waterways, but oil companies build roads into the forest, which are later used by poor farmers as access to clear more land.

A few years ago I was helicoptered into Vilcabamba, a rainforest in a mountainous region of Peru, a part of the Andes, between its peak and the Amazon Basin. The purpose of the expedition was to explore the region and survey its species. It was sponsored by Conservation International, a Washington, D.C.–based environmental group.

The Shell, Mobil, and Chevron corporations had won concessions from the Peruvian government to explore the valleys just east of Vilcabamba and to build a pipeline to Lima over the Andes. Under this eminent threat, Conservation International decided to send in their RAP (Rapid Assessment Program) team, a group of American and Peruvian biological superstars (biologists from the Smithsonian Institution, the Field Museum of Natural History in Chicago, and the Museo de Historia Natural in Lima, Peru). The goal was to survey the wildlife of Vilcabamba in four weeks. Scientists believe the tropical Andes are among the most biologically diverse of all the remaining forests in the world. Vilcabamba, an uninhabited wilderness in the midst of this range, roughly the size of New Hampshire, has enormous conservation potential.

It rained every day during my stay in the Vilcabamba rain forest. In the morning, there were brief periods of sunlight during which the biologists would hang wet clothing on every available tree, tent, or rope in desperate hope of drying them out. They'd pull them in quickly when the clouds started to build. The forest was so dense you couldn't see that many animals until the biologists put out nets that would quickly fill with tropical birds by day and bats by night, which biologists would catalogue and release. They also spotted white-faced capuchin monkeys, spider monkeys, and night monkeys. And they found signs of puma, dwarf deer, and bear.

The biologists discovered a number of plants, animals, frogs, and butterflies that had never been recorded before. Tom Schulenberg, the RAP team leader, was the local bird expert. In the weeks that we were there, he worked the edge of the bog in the middle of the forest, training both binoculars and microphones into the trees. According to Schulenberg you can hear approximately three to four times more birds than you can see in the rainforest, and he can recognize the calls of different birds. He believes that Vilcabamba has at least 650 species of birds, roughly equivalent to the total number of species that breed in the United States and Canada combined.

Mobil, Shell, and Chevron are making major efforts to promote their new ecological awareness, and scientists are working with them. But they are cautious since Mobil made such a show in Tambopata, a mecca for bird watchers and a proposed national park in southeast Peru. At the last minute, Mobil walked off with a concession to explore the heart of the park, and conservationists got only a third of the space originally proposed.

> **Climatoids**
>
> The tropical Andes is what scientists refer to as a "hot spot," a term coined to identify areas with large numbers of species, found nowhere else on Earth, that are concentrated in a relatively small space. It's thought that if we can at least save these areas, we will be able to preserve a maximum number of species given the limited amount of conservation dollars.

Slash and Burn

Many rainforest farmers practice slash-and-burn agriculture. When they cut and burn the forest, it releases nitrogen and other fertilizers held by the trees back into the soil. Farms cleared this way are usually productive for a few years but then the soils are exhausted, and the farmers abandon their farms and clear more tracts. If allowed to lie long enough, cleared forest like this will rejuvenate, but the problem is that competition for arable land is so intense, the land is not allowed to rejuvenate, and farmers cut down more and more forest.

Brazilian scientists, monitoring the rainforest by satellite, estimate there are more than eight thousand fires in a single day in the Amazon rainforest. For the last two decades the annual emissions of carbon dioxide from the burning of forests has equaled some of the world's worst polluters. Brazil ranks with the United States, Russia, and China in greenhouse gas pollution, yet it has little industry. Without the burning, Brazil would not even make the top twenty polluters.

A Double-Burger and a Side of Rainforest

Cattle ranching takes its share of the rainforest in Central and South America. Ranchers clear the rainforest and plant pasture for cattle. However, as with rainforest farming, the soils are soon exhausted and more forest must be cleared. During the 1970s and 1980s, much of the beef raised on these rainforest cattle ranches was shipped to Canada and the United States to be used as hamburger for fast food restaurants. In the 1990s both McDonald's and Burger King agreed not to buy beef from lands that were previously rainforest.

According to a 1997 report from the Food and Agriculture Organization of the United Nations, the transformation from tropical forest to crop and pastureland brings about substantial losses of soil fertility and soil erosion. Pastures cleared from rainforest can only be sustained for about 10 years. More than 50 percent of the pasture areas in the Amazon rainforest have now been abandoned in a degraded state.

The Least You Need to Know

- ◆ Forests once covered most of the earth but have been reduced by as much as 50 percent.

- ◆ Industrialization and the clearing of land for agriculture depleted the temperate forests of North America and Eurasia.

- ◆ The rainforests of the world make up a diverse storehouse of plants and animals.

- ◆ The rainforests are disappearing at the rate of two football fields (about 4 acres) every second.

Part 3

A Meteorological Primer

If you want to talk about global warming, you have to understand the weather—at least a little. In this section, I'm going to bring you up to speed. I'll tell you how winds are formed, how rain is made, and how hurricanes are born. I'll tell you the difference between weather and climate. (Hint: Weather is what we expect over the next 10 days. Climate is what we expect over the next 10 years.)

I'll talk about El Niño. What can a few added degrees of warmth do to the weather? El Niño is a good example. I'll take a look at some of the other climate variables that could be creating the warming we're experiencing. What if it's just sunspots? Wouldn't that tick the scientists off?

Lastly, I'll take a look at some of the variables—like deep ocean current, the permafrost, and melting polar ice—that could throw the whole atmospheric system out of whack and ratchet up the global-warming equation.

Weather 101

In This Chapter

◆ Where we get clouds

◆ The lows and highs of atmospheric pressure

◆ From drizzles to hurricanes

◆ Global air circulation

When we talk about global warming and its effects we are essentially talking about how increased temperatures worldwide might change the weather and the climate. *Weather* is the condition produced by the temperature, humidity, wind, and precipitation at a given moment. *Climate* represents the weather averaged over a number of years. When we say the climate is getting warmer, we are not talking about the weather on a given day or a given place, we're talking about the weather averaged out over the entire world and number of years, decades even. Because you were shivering last night in Alberta or sweating midday in Death Valley doesn't disprove global warming. When we are talking global warming, we are talking averages or climate.

Still climate will affect the weather. Changes in globally averaged temperatures will have effects on rain, snow, sleet, and hail, as well as tornadoes and hurricanes. It will affect which areas get wetter and which ones get drier. To understand more about the bigger climate picture, let's take a look at those things that affect the weather right now.

> ## Warm Words
>
> **Weather** is the current condition produced by temperature, humidity, wind, and precipitation. **Climate** refers to an average of those conditions over a number of years.

Temperature and Humidity

Good old water, H_2O, exists in three different states: as a solid, a liquid, and vapor. The temperature of the air determines in which state you'll find it. When the temperature falls below 32°F (0°C), water molecules stop moving freely, bind together, get hard, and turn into ice, snow, sleet, and hail. When it gets hot, water turns to vapor. The molecules start moving freely, drift apart, and become clouds, fog, or humidity.

> ## Climatoids
>
> During the winter months the ability of the atmosphere to retain water vapor is almost four times higher in the tropics than it is in the mid-latitudes because of the differences in temperatures.

> ## Warm Words
>
> **Relative humidity** is a measure of the air's capacity to hold moisture versus the actual moisture present. To get rain you have to have 100 percent humidity.

Much of the water in the atmosphere is held as water vapor and is therefore invisible. If the temperature is warm, the air has a greater capacity to hold water as vapor. If that capacity is reached, the excess water will fall out in some form of precipitation. Two things can create precipitation: Either more water vapor is added to the air or the temperature drops, which decreases its capacity to hold that vapor.

One important term you might hear when listening to the weatherman is *relative humidity*. This is the ratio between the actual amount of water vapor in the air and its capacity to hold that vapor. The relative humidity of the air increases if more vapor collects in the atmosphere or if the temperature is lowered. When it heats up, the relative humidity goes down. When the temperature falls, the relative humidity goes up. When relative humidity reaches 100 percent, it rains, snows, hails, or does something else to spoil your picnic.

Clouds

But getting from a state of 100 percent relative humidity to rain or some other type of precipitation isn't that easy. It has to first form into clouds, and there are a lot of complex processes involved. (Why else would we have the Weather Channel?) In fact, the water vapor in the air has a difficult time coalescing into cloud droplets or rain-drops. The surface tension around the tiny vapor droplets is fairly strong. By them-selves, water vapor molecules are not attracted to each other enough to form cloud droplets or rain droplets.

That's where those particulates in the atmosphere (that I mentioned in Chapter 2) come into play. Their role in the water vapor-to-raindrop problem is to act as *conden-sation nuclei* around which cloud droplets can form. These particles come from ocean spray, forest fires, volcanoes, and smoke stacks. That's why it rains more around pol-luted factories. Even when the atmosphere looks crystal clear, there are still plenty of these microscopic particles in the air. In a volume no larger than a finger, there could be over 100,000 of these particles.

There are different kinds of clouds. The four basic types are *cirrus, stratus, cumulus,* and *nim-bus*. Cirrus clouds are high, thin, and wispy. The high altitude and cold creates ice crystals which cause the wispy effect. Stratus clouds are flat and are found closer to the ground. They like to drizzle. Cumulus clouds are puffy. Nimbus clouds are any clouds that deliver rain. (Cumulonimbus clouds are cumulous clouds that contain rain.)

> **Warm Words**
>
> **Condensation nuclei** are microscopic particles in the air around which cloud droplets and raindrops form. **Cirrus** clouds are high and wispy. **Stratus** clouds are flat and low to the ground. **Cumulus** clouds are puffy and unthreatening, or gathering, ris-ing, and thunderous. **Nimbus** clouds are rain-bearing clouds.

The elevation at which these clouds arrive is important. High clouds, over 20,000 feet, may foretell the arrival of a storm. Middle clouds, from 6,500 feet to 20,000, are thicker. When high clouds give way to middle clouds, rain is approaching. Low clouds form at elevations below 6,500 feet and come filled with rain or snow. But low clouds generally deliver their precipitation steadily, not as thundershowers. Clouds that rise vertically, frequently into thunderheads, occur throughout the various levels. They are associated with strong upward currents. They have a puffy, even towerlike structure. Precipitation is heavy but not that long-lasting. Generally, the bigger they grow, the more violent the weather becomes.

Precipitation Wears Many Hats

Precipitation comes in a variety of forms as shown in the following list. Temperature is important, as well as the upward movement of the air.

◆ **Rain** Rain is formed when cloud droplets coalesce. Over the ocean the salt in the air helps the cloud droplets bind together, but over land, ice and microscopic particles (particulates) do the job.

◆ **Snow** Snow develops when ice crystals link to each other. Although snow is beautiful, it plays havoc with urban life. Lines form at the grocery store when the weatherman forecasts snow. Snow comes in an infinite variety of crystalline shapes.

◆ **Sleet** Sleet is frozen rain that occurs in winter when warm air rises over a cold layer. Rain falls from the upper warmer layer through the cold lower layer, which is not quite cold enough to cause snow but cold enough to convert the rain to frozen pellets.

◆ **Hail** Hail is more of a warm weather phenomenon. Hail forms when ice falls through clouds with vertical development but which have updrafts that keep sending the hail stones aloft, creating new layers of ice until they get so heavy they fall from the sky.

◆ **Freezing Rain** Freezing rain turns to ice upon contact with cold surfaces. These rains are usually caused by ground-level freezing winds. This creates a glaze of ice that coats roads and walkways.

Why It Rains

Getting from water vapor to rain is complicated. First the air has to be at approximately 100 percent relative humidity. We get there by adding more moisture or reducing the temperature. But it turns out that just bringing in more water won't get the microscopic droplets to bind together, we need a drop in temperature.

Climatoids

Rising air produces clouds and rain, but falling air produces deserts. Such are the conditions on the West Coast when winds off the Pacific hit the Sierra Nevada Mountains. Rising air creates precipitation, which is why the western half of the Sierras is covered with trees and lakes. East of the Sierras is Death Valley. The same situation exists in Washington and Oregon, where rain rising over the Cascades creates a densely forested area along the coast, but a desert inland.

The surest way to lower the temperature is upward movement of the air. Remember from Chapter 2, it gets colder as we go upward in the lower atmosphere. As it gets colder, the relative humidity gets higher. At 100 percent, clouds form. If it gets high enough, ice forms, even during the summer, even in the tropics. Rising air is your best bet for rain or any of its relatives.

The upward velocity of the air affects how the moisture will develop. If the air is rising slowly, it may only form into clouds or light drizzle. If it's going up fast it might fall out as heavy rain or a torrential downpour. Dense, rapidly building clouds are the stuff of thunderstorms and tornadoes.

Under Pressure

Rising air is also where the low-pressure areas are, which makes a nice segue into the next topic—pressure. If you are feeling a little pressed lately, it might be the atmosphere. Actually 10 to 20 tons of atmosphere rests on the average adult. The reason we don't all collapse is that our bodies exert an equal amount of pressure from within. (Bet you didn't know you were that strong.) We actually go through life with this load on our backs and don't even notice it. Pressure is important in weather, though.

We measure pressure with a *barometer*. The barometer measures pressure in inches (millibars) of mercury. Normal sea-level pressure is 29.92 inches (1013.2 millibars). At ground level, atmospheric pressure fluctuates only a few percent, generally between 30.50 and 29.50 inches. Because the air rises when it rains, the pressure will be lower. But severe storms will seldom have a pressure lower than 29 inches. Hurricanes can have a pressure lower than 28 inches. Higher elevations have lower pressures because there is less air bearing down from above. Meteorologists speculate that the lowest pressure would be within a tornado, but nobody has figured out how to get a barometer to survive a tornado.

Warm Words

A **barometer** is a gauge used to measure the pressure of the atmosphere.

Low Pressure

Low pressure is where the action is—the storm action, that is. In low-pressure areas the air is rising, which creates conditions for rain and its related wet friends. *Converging* surface winds can cause low pressure. The winds come together and there's no place else to go but up. Upward motion creates lower pressures because the upward-moving air is pushing against the weight of the atmosphere. *Diverging* currents in the upper atmosphere can also cause

Warm Words

Converging currents in the atmosphere occur where winds meet or where more air enters an area than departs. **Diverging currents** occur when winds depart in different directions or when more air leaves than arrives.

lower pressure, since air moving out from a given point in the upper atmosphere will cause air from the lower atmosphere to be sucked upward to fill the vacuum. Low pressure, therefore, is all about rising air currents.

High Pressure

High pressure is where the picnic and beach weather is. In high-pressure areas the air is sinking and as the air sinks it gets warmer. Diverging surface winds can cause high pressure. The winds move out from a given surface point pulling the air above down into the void. Downward motion creates high pressure because downward-moving air is adding to the weight of the atmosphere. Converging currents in the upper atmosphere will cause the air meeting at the point of convergence to move downward. High pressure is all about sinking air.

Storm Fronts

Storm fronts occur where air masses meet. Air masses are enormous bodies of air that have pretty much uniform moisture and temperature. A single air mass can cover one half of the United States. Air masses are given names based on the region in which they were born. Thus, air masses can be arctic, polar, tropical, or equatorial, depending upon the direction from which they come and either maritime or continental depending on whether they were born out over the ocean (maritime) or the land (continental). Thus a maritime polar air mass might form off the east or west coast of Canada and a maritime tropical air mass might form off Mexico. A continental polar air mass might form in the middle of Canada; a continental tropical in the middle of Mexico.

Arctic and polar air masses are cold; tropical and equatorial air masses are warm. Maritime air masses are moist, and continental air masses are dry.

When these masses meet, you get a front. And fronts are trouble. They are often where the storms occur. You can have cold fronts or warm fronts. A cold front is where colder, drier air overtakes warmer, moister air. Cold fronts create the most violent conditions. Cold fronts move faster and have a steeper face. The cold, heavy air lifts the warm, lighter air rapidly. As the warmer, wetter air rises, the relative humidity increases, tall clouds form, and storms happen. Again, there is vertical motion of the air, which is the key to storm action.

Warm fronts occur when a warm air mass moves up on a cold one. Warm air masses don't move as rapidly. They slide up over the cold air masses more gently, at a more gradual angle. The clouds will be lower, the precipitation less intense. Sometimes all you get is drizzle or fog.

A storm approaches.

(Photo by Michael Tennesen)

Doing the Cyclone

Frontal cyclones are those big circular cloud formations you see from the weather satellites that everybody is always getting so upset about. The *Coriolis* effect creates the spinning. The Coriolis effect causes the wind to veer to the right in the northern hemisphere and to the left in the southern hemisphere. The shift is due to the fact that air at the equator is moving faster than it does at the poles. *Frontal cyclones* generally occur at the junction of two fronts. The spin adds something extra to the front. To get your really big storms you have to have diverging winds in the upper atmosphere which act as an exhaust fan, drawing the air up from below. Take a low-pressure zone, put some spin on it, add an exhaust fan, and you've got the makings of a whopper.

Warm Words

The **Coriolis** effect causes winds to veer to the right in the northern hemisphere and to the left in the southern hemisphere. **Frontal cyclones** are large spinning weather systems that occur along turbulent fronts.

Tornadoes

More tornadoes occur in the United States than in any other country in the world. That's because the United States is caught between the cooler air masses of Canada and the warm, tropical ones further south. Spring is the best time for tornadoes. That's when cold air masses still linger over the continent as warmer tropical air is moving up fast through Texas and the lower plains.

The lighter, warmer air rushes up over the lingering cold mass. A funnel can develop like a rolling log at the junction where the masses brush past each other. A strong downdraft can upend that spinning log and a tornado can touch down.

Tornadoes can tear down a brick wall. A severe tornado can pick a car up and toss it in the air. Tornadoes destroy instruments that measure the wind, but researchers have gotten around that by using Doppler radar to estimate speeds, and tornado winds can exceed 300 mph (480 kph).

> **Global Warnings**
>
> Most tornadoes in the United States occur in the region known as "Tornado Alley" that extends through the central plains states. There are 600 to 700 tornadoes reported each year in the United States. The most costly tornado incident was a cluster of tornadoes that ran through Iowa, Illinois, Wisconsin, Indiana, Michigan, and Ohio on April 11, 1965. It killed 271 people and caused $300 million in property damage. The most deadly tornado was a single one that ran through Missouri, Illinois, and Indiana on March 18, 1925. It killed 689 people. One of the effects of global warming could be an increased amount of tornadoes.

Hurricanes

Hurricanes are the mean guys, the worst storms we've got. Hurricanes are called typhoons in the western Pacific and cyclones in the Indian Ocean. They're all the same animal. Hurricanes are tropical cyclones whose winds exceed 73 mph. Unlike frontal cyclones, hurricanes don't occur along fronts of contrasting air masses. There aren't any big contrasts in the tropics; temperatures are all about the same. Hurricanes don't run on contrasting temperatures, they run on heat.

Most hurricanes that strike the eastern seaboard of North America originate off the northwest corner of Africa and move west over the Atlantic. They start with a cluster of thunderstorms converging. Remember surface convergence creates low pressure because the winds are sent upward. Tropical heat assists the rise. Hurricanes form over oceans where the water temperature is at least 79°F.

As with frontal cyclones, they need that exhaust fan—the high altitude divergence of air—to really get things going. Hurricanes are some of the few storm systems which might actually interact with the stratosphere. Hurricanes can cover an area with a diameter of from 300 to 700 miles. A hurricane has spiral bands that radiate out from the center. Most of the rain occurs within those bands. The winds gradually increase and reach their greatest intensity in the eye wall, the area surrounding the eye of the hurricane. Some scientists believe that small tornadoes form in that eyewall and do the damage. The eye of the hurricane itself is relatively calm.

Global Warnings

Hurricanes work their destruction by violent winds (Hurricane Andrew [1992], $40 billion in damage), torrential rains (Hurricane Mitch [1998], 11,000 dead), and enormous storm surges of water (Galveston Hurricane [1900], 6,000 dead.)

Sun Matters

Weather isn't all about water and wind—the sun matters. The sun's heat collects in differing amounts around the surface of the globe and this is the great cause of weather. The tropical latitudes get lots of heat, while the polar regions get little. The atmosphere and the ocean then try to even the heat out. Some of the equator's heat moves toward the poles, and some of the pole's cold moves toward the equator, trying to achieve a balance. Without these motions the tropics would be even hotter, and the poles even colder. As the air moves back and forth, storms form, sea breezes begin to blow, rain falls, and hurricanes develop.

Global Circulation

Although winds can change on a daily basis, there are some consistencies in world wind patterns. Working from the equator toward the poles, there are three bands of wind directions in each hemisphere. From the equator to about 30° latitude (about a third of the way to the poles), the winds come from the east and are called the trade winds. Actually, they come from the northeast in the northern hemisphere and from the southeast in the southern hemisphere. In the middle latitudes, from about 30° to 60°, they come from the west and are called the prevailing westerlies. And from the poles, they come from the east and are called the polar easterlies. The Coriolis force causes them to arrive at slight angles. Check out the following figure to get an idea of which way the winds blow.

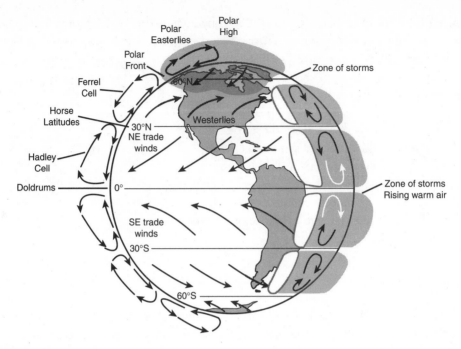

How the atmosphere circulates.

What regulates these winds are three distinctive cells or circular wind patterns in both the northern and southern hemispheres—one near the equator and one at the poles, and one in between. The trade winds from both hemispheres converge on the surface at the equator, which sends the air aloft, reducing the pressure. This is why it rains so much around the equator. When the air reaches the upper altitudes, some of it heads north, and some heads south in an effort to even out the heat difference between the equator and the poles.

At about 30° latitude, the high-altitude winds sink to the surface. Sinking winds create high-pressure zones, with lots of sun, which is why 30° latitude is where the world's great deserts are. Some of that sinking air heads back toward the equator, which keeps this cell or circular wind pattern going. See the previous figure.

Some of the high winds descend at the poles. From there they head back toward the equator and converge with the prevailing westerlies at 60°. When winds converge, they rise. Thus, you get another area of frequent storms at 60°.

In general, there are a number of prevailing wind conditions on the surface and at higher altitudes in both hemispheres that greatly affect the weather. Where winds converge on the surface at the equator, the air moves upward and you will have lots of storms and rain. At 30° (one third of the way toward the poles) winds diverge at the surface, high winds move downward, and you get deserts and dry weather. At 60° latitude (two thirds of the way to the poles from the equator) you get converging winds and another uplifting with resultant big storm activity.

> ### Climatoids
>
> At 30° latitude the surface winds diverge and air from upper altitudes must descend to fill the void. The descending air creates a high-pressure zone that's filled with sun, dry air, and not much wind. Over the oceans this is called the horse latitudes. This is because at this latitude, sailing ships of old would encounter calm and the ships wouldn't budge. Sailors then threw everything overboard to lighten the load—including the horses.

Though surface winds vary at different latitudes, above 15,000 feet (4,560 meters) in the troposphere (the bottom layer of the atmosphere), the wind is generally from the west. This is why the big storms move from the west to the east. During the summer the winds can shift in the tropics, which is why hurricanes head east.

So we have this complex system, where air moves in complex patterns, stirring up trouble in the process. Now what are a few extra globally warm degrees going to do? Well, a little extra heat is all it takes to get an El Niño going. We'll look at what that can do in the next chapter.

The Least You Need to Know

- Moisture and humidity go through changes to get to rain.

- High-pressure zones are sunny and calm; low-pressure zones are cloudy and stormy.

- Rising air creates low-pressure zones; sinking air creates high-pressure zones.

- Where air masses with different temperatures and moisture content collide, look out for storms.

- The air over the globe moves in complex patterns created by the movement of air from the equator toward the poles and complicated by the rotation of the earth.

Chapter 11

El Niño

In This Chapter

- The incredible force of warm ocean water
- Mudslides in California
- Ice storms in New England
- El Niño ups its ante

Gary Lemonoff will never forget the night during the El Niño storms of March 1998, when the rain-saturated earth above his home near Laguna Beach, California, gave way, and tons of earth, rock, and trees slammed into his house. In the blackness of the night, he could hear his neighbors screaming. But the real horror awaited him in the early morning light when he looked out his window to see an arm and a leg protruding from the mud and rubble. Two people died in that slide.

But this was just a small portion of the devastation brought by the 1997–1998 El Niño, one of the most powerful climate events of the century. The United Nations estimates that massive storms, fires, floods, frosts, and droughts killed 2,100 people and caused in excess of $33 billion in property damage. El Niño has been with us for a long time, but its potential to disrupt the weather worldwide was not systematically studied until the International Geophysical Year of 1957–1958. From this we've learned something of how small our planet is and how interrelated is its weather.

What's Another Five or Six Degrees?

El Niño is a warm body of water that appears every two to seven years around Christmas time in the Pacific Ocean off the coasts of Peru and Ecuador. It usually disappears by the end of March but can persist up to 18 months. The increase in temperature isn't much: 2–3.5°F (1–2°C) in most years, 7–11°F (4–6°) in strong El Niños. What's interesting is how close this is to the IPCC's prediction of global warming, which they estimate at 2.5–10.4°F (1.4–5.8°C) over the next century. El Niño, in some ways, provides us a working model of how additional heat might affect global weather.

> **Climatoids**
>
> El Niño is Spanish for "The Christ Child," a term used by Ecuadorian and Peruvian fishermen for the warm ocean phenomenon that normally appeared around Christmas.

How El Niño Works

Remember those easterly trade winds that are found around the equator, which I mentioned in Chapter 10? Well, in an El Niño year, those winds weaken, allowing warm water in the eastern Pacific around Australia and Indonesia to flow eastward toward Peru and Ecuador. Normally the trade winds move that warm water eastward, which causes *upwellings*, the upward movement of deeper, colder water to the surface off South America. That colder water is richer in nutrients. Those nutrients feed the plankton, which feeds the fish, which keeps the big fish, the seals, and the fishermen happy.

In an El Niño, the warm water comes and buries those cold upwellings under about 500 feet of warm water, pushing the *thermocline*, the boundary between the warm water and the cold water, lower. In normal years the thermocline is only about 130 feet below the surface. The area of warm water is huge—about one and a half times the size of the United States. The 1997–1998 El Niño bore more energy than a million Hiroshima bombs. Its energy spawned a whole suite of other weather phenomena, including floods, droughts, tornadoes, and hurricanes.

> **Warm Words**
>
> **Upwellings** are the upward movement of deeper water to the surface. The **thermocline** is the boundary between the ocean's warm and cold waters.

By July and August of 1998, El Niño had begun to fade, but not completely. June 1998 was still the warmest June in the historical record. El Niño helped make 1998 the warmest year in the past 140 years and probably in the last 1,000.

The History and Effects of El Niño

The effects of El Niño were documented in Peru as early as the Spanish conquest in 1525 C.E., but researchers have found geological evidence of El Niño going back 13,000 years. Anthropologists believe that the Incas were aware of El Niño. That's why they built their homes on the tops of hills. If they did build near the coast, they kept back from the shore and stayed away from rivers. By the end of the nineteenth century, Peruvian oceanographers were studying the phenomenon. At the time they thought El Niño's effects were limited to the west coast of South America. It wasn't until the systematic studies of the International Geophysical Year that the global extent of its impact was recognized.

In normal years in the equatorial regions of the Pacific, sunlight heats up the upper layer of seawater around Australia and Indonesia. When it gets hot enough, the water begins to evaporate, and the hot, moist air rises, creating a low-pressure zone and lots of rain. It's part of the monsoons that hit south-east Asia. When the air gets high enough, the westward winds of the upper troposphere send it toward South America. By the time the air hits the Latin coast, the winds are colder and dryer, and so they sink. The sinking air creates a high-pressure zone, which is why there is dry desert along the west coast of South America despite its proximity to the equator.

> **Climatoids**
>
> The giant circulatory loop that moves in the upper atmosphere from Australia and Indonesia to South America and then back along the surface is called the Walker circulation for Sir Gilbert Walker, the British scientist who first studied it in the 1920s.

The sinking high-altitude air hits the shoreline where the normal surface westward-moving trade winds push it back toward the western Pacific. As these trade winds blow westward over the Pacific, they push the warm top layer of water with them. Along the coast of South America, but especially off the coast of Ecuador and Peru, colder subsurface water wells up to replace the departing warm water, bringing up a bevy of nutrients for all the creatures in the sea. The end result is that the coast off Peru is one of the most prolific fisheries in the world.

It Kills the Fish

But during an El Niño, the trade winds diminish or disappear. The warm water in the eastern Pacific just sits off South America soaking up the heat, blocking the cold upwellings, along with the feast of nutrients they provide for marine life. Fish that normally thrive off Ecuador and Peru head south in search of cooler, richer waters.

The fishing gets good in Chile. The change is felt by marine life as far north as the Canadian border. During the 1997–1998 event, dorado (mahi mahi), a fish normally caught in tropical waters, were caught by sport boats in the waters off California. One fisherman caught a marlin off Washington State, and marlin are rarely seen north of Mexico.

When an El Niño arrives, Peruvian fishermen usually take a break, fix their nets, repair their boats and spend time with their families. Mild events are generally tolerated; strong events can be economically devastating. The 1997–1998 El Niño put a serious dent in the Peruvian fishing economy.

El Niño affects fishing.

(Photo by Michael Tennesen)

It Floods the Cemeteries

As the ocean gets warmer, the water off Ecuador and Peru actually swells. During normal years, because of the westward winds, which push the water with them, the sea level is usually 18 inches higher off Indonesia than it is off South America. During an El Niño the situation is reversed. As the water off South America gets hotter, it starts to evaporate. The moisture moves upward, creating a low-pressure zone off Ecuador and Peru. Torrential rains begin to fall in areas that aren't used to it.

In Peru during El Niño 1997–1998, mudslides, fed by the second-worst El Niño rains in Peruvian history and aggravated by loggers that clear-cut the forests of nearby hills, destroyed the entire village of Santa Teresa, killing 22 people. Three hundred bridges were washed out. To cross the San Francisco River many of the people from surrounding towns had to commute by inner tube pulled by the local ferrymen. Flooding in the town of Mampuesto unearthed caskets from the local cemetery and floated skeletons through the streets.

El Niño's Global Reach

Though El Niño involves only one-fifth of the surface of the planet it transforms weather around the world. El Niño currents not only influence the water and local weather, the additional heat in the eastern Pacific feeds into the high atmosphere and creates a strong jet stream that races northward. It's always there, but it's much more powerful during an El Niño. The jet stream turns onto the North American continent and marches across to the Atlantic. Its effects are enormous.

> **CAUTION**
>
> **Global Warnings**
>
> The runoff from the flooding in Peru during 1997–1998 was so severe that it created a lake in the coastal desert, where there had been only arid wasteland before. The lake, 90 miles long, 20 miles wide, and 10 feet deep became for a time the second-largest lake in Peru. It served as a warning of the climate's ability to alter the landscape on a grand scale.

Mudslides in California

California seems to get the brunt of the El Niño storms delivered to the United States. During the 1997–1998 El Niño, the local media kept doing stories on its coming impact, but as fall turned to winter and little rain fell, El Niño became the butt of nightly jokes on the weather report. But around February, the rains began to fall and the jokes stopped.

The first 24 days of February, Los Angeles received 13.5 inches of rain, close to its annual average, and the most since records were first kept in 1877. San Francisco's total for the season was 38.5 inches, making it the wettest winter of the century. These totals don't sound like much to someone from the Northwest or the Northeast. But you have to consider that the vegetation on the hills around California have evolved to hold much lower amounts of water. The rain saturated the ground and quickly exceeded the ability of the sparse vegetation to hold it back.

Much of the destruction created in California by El Niño came as mountains and coastline cliffs simply seemed to dissolve into debris slides or mudslides, taking homes, cars, livelihoods, and lives with them. The Amtrak rail line between Los Angeles and Seattle was severed by a mudslide in Santa Barbara. Ocean waves reached heights of 30 feet in spots along the coast. Helicopters regularly pulled surfers from the water. The capacity of the normal drainage systems, overflow channels, and riverbeds was quickly exceeded and excess water became destructive. Two highway patrolmen died before dawn when their car was swept into a swollen river east of Santa Maria. Two college students were killed when a tree fell on their vehicle.

I remember being caught in a torrential downpour on the Los Angeles Freeway during one of the more severe of the season's downpours. Visibility was suddenly cut to about 10 feet in the middle of the day, as motorists slowed from 50 miles per hour to 5 miles per hour. All of us braced for the possible motorist who hadn't bothered to brake or whose reflexes were unused to the conditions to plow into the rear of one of our cars and set up a chain reaction.

Hurricanes Off Mexico

The warm, eastern Pacific waters set up prime conditions for hurricanes. Hurricane Linda grew out of that eastern Pacific in September 1997. Her 185-mile-an-hour winds made it one of the strongest eastern Pacific storms ever recorded. The spiral arms of the hurricane lashed Mexico's western port cities. I took a bus down the coast from Puerto Escondido to Santa Cruz along the southern coast in the state of Oaxaca, Mexico, in the fall of 1998, a year after the hurricane, searching for sea turtles. I was amazed at how many concrete structures had been demolished by the storm. The sea turtle center at Puerto Angel, one of Mexico's prize natural exhibits, lay in ruins.

Although the eastern Pacific delivered up a number of intense hurricanes in 1997, the same was not true for the Atlantic. An El Niño-related tropical jet stream headed eastward across the Atlantic, effectively sheering the tops off the clouds before Atlantic hurricanes could form. During 1997, tropical storm activity in the Atlantic was diminished. The month of August passed without a single tropical storm developing over the Atlantic, the first time that had ever happened in the twentieth century.

Warm Words

Jet streams are strong ribbons of horizontal winds that are found about six to ten miles above the ground in the area between the troposphere, the lower layer of the atmosphere, and the stratosphere above it.

Cherry Blossoms in D.C.

While the Southwest endured one of the wettest winters in recorded history, the winter in the Northeast was the warmest on record. Cherry blossoms were blooming in January in Washington, D.C. The jet streams that travel six to ten miles above the earth's surface shifted dramatically. The polar *jet stream* stayed further north over Canada than usual. The typical winter chill was reserved for those north of the U.S. border. It was a year of no winter for the mid-Atlantic region. Some estimates put the savings in heating oil for that year at five billion dollars.

Ice Storms in New England

But winter didn't disappear, it just shifted northward. Quebec and northern New England were bashed with one of the worst ice storms in their history. Ice storms, you remember, occur when warm air flows over a freezing surface. Southern New England to northern Georgia are the normal receptors of this type of storm. But during the big El Niño winter, the pattern moved northward and a heavy layer of ice accumulated during a storm in January. Ice felled enormous power lines and trees. Quebec and northern New England were brought to a standstill. It was one of the biggest ice storms ever to strike the East in recorded history.

Tornadoes and Floods in the South

Vicious storms pummeled the southern states. Snow fell in Guadalajara, Mexico, in the winter of 1997–1998 for the first time since 1881. In December 1997, Roswell, New Mexico, experienced its worst one-month snowfall since 1893. From Florida to Texas, tornadoes and floods were relentless. During the late evening of February 22 and early morning of February 23, 1998, a series of tornadoes ripped across central Florida. At least one of the tornadoes reached an estimated rotational wind speed of 260 miles per hour at the center. The tornadoes destroyed 800 residences, and left another 700 uninhabitable. Damages exceeded $60 million. Forty-two people were killed, making it the deadliest outbreak of tornadoes in the state's history.

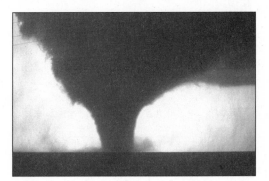

El Niño spawns tornadoes.

(Photo courtesy of FEMA)

Droughts in Asia

El Niño moves the rains that would normally fall on Indonesia, Australia, and India to the eastern Pacific and produces drought in the western Pacific. According to historical records, 600,000 people died in India from the epic droughts that accompanied the El Niño of 1789–1793.

Forest Fires in Sumatra

During the 1997–1998 El Niño, forest fires burned furiously in Sumatra, Borneo, and Malaysia, forcing drivers to use their headlights at midday. The smoke traveled thousands of miles to the west into the normally crystal clean air of the Maldive Islands, where the visibility was limited to half a mile at times.

The Fall of the British Empire

One historian argues that El Niño could have contributed to the fall of the British Empire. According to Mike Davis in his book, *Late Victorian Holocausts: El Niño Famines and the Making of the Third World* (Verso Books 2001), the oscillations of the giant Pacific current and the revolutions of the early twentieth century are linked. Davis recounts how in the late nineteenth and early twentieth centuries a number of El Niño–related events led to famines in India, Brazil, and China. But famines were so ineptly handled by British imperialists that, according to Mr. Davis, they resulted in the deaths of 50 million people. Davis argues that imperialism had already robbed these formerly prosperous areas of their resources, and when calamity hit, the people had nothing to fall back onto. When the imperial government didn't fill the void, its fall from power began.

> **Global Warnings**
>
> Asia, Indonesia, and Japan were hit by famine and fire during the 1997–1998 El Niño, to the extent that it stretched thin the economic safety nets that many of these countries had during the late '90s. Though it was certainly not the only cause, many economists believe it was one reason for the great Asian Stock Market Crash of 1997. If global warming predictions are correct, what economic hardships might a rash of such climate-related events produce in the twenty-first century?

El Niño and Nature

The warm water buildup off Ecuador and Peru is fatal not only to the fish but to other marine life as well. In 1982–1983 during an El Niño, 25 percent of the adult seal and sea lion population perished, as did most of their pups along the coast of Peru. During an El Niño event, the warmer-than-normal water sends tropical fish northward, but off the west coast of the United States, those fish don't take up the void created by the disappearance of normal prey species.

El Niño plagues California sea lions.

(Photo by Michael Tennesen)

The 1997–1998 El Niño devastated the California sea lion population in the islands off California and Baja California. Nursing females usually find plentiful squid and small fish near the surface, but the prey swam deep in search of cooler waters, out of sea lion diving range. In non-El Niño years pups have a rough time, with mortality usually reaching 25 percent. But in 1997–1998, pup mortality reached 70 percent.

The heat that cooked the waters off Australia took a toll on marine life there as well. Along Australia's Great Barrier Reef, the water temperature got as high as 88°F (31°C). The coral polyps along some sections of the reef were bleached white. Most areas recovered, but along an 18-mile stretch off western Australia, the coral eventually died.

El Niño and Disease

El Niño brought more water to Peru and that brought more disease. Water pooled in the eastern desert, and mosquitoes thrived. Those mosquitoes were responsible for the spread of malaria, some 30,000 cases in the Piura region alone, three times the average.

Bartonelosis, a bacterial disease transmitted by the bites of sand fleas, rampaged through cities

Warm Words

Bartonelosis is a bacterial disease transmitted by sand fleas that causes wart-like sores on the skin in milder cases and anemia in more severe cases. **Anemia** is a disease in which body tissues are deprived of oxygen by a reduction in the number of red blood cells or inadequate amounts of an essential protein called hemoglobin.

on the western slopes of the Andes. In the chronic form of the disease, patients get long-lasting, blood-filled, wart-like sores on or under their skin. In its acute form, the disease can cause *anemia* and is fatal in as many as 40 percent of untreated cases.

Residents of Cuzco, a popular tourist town in the Andes near the Incan ruins of Machu Picchu, had never suffered outbreaks of bartonelosis until the 1997–1998 El Niño. Data from satellites that monitor the temperature of the Pacific show that the ocean began to warm about two to three months before the disease hit.

These weren't the only problems. Between January and May of 1998, 1,039 clinical cases of cholera were reported in Lima, Peru. The outbreak of that disease also coincided with the increase in temperature. The peaks in temperature preceded the outbreak by three weeks.

La Niña, the Wicked Sister

La Niña is the wicked sister of El Niño. During a La Niña, an abnormal cooling in the eastern Pacific produces conditions more or less the opposite of El Niño. As with El Niño, the effects are most pronounced from December through March. During a La Niña, westbound trade winds are intense. That drives more than the normal amount of warm surface water westward. More than the normal amount of deep chilly water rises up off the coast of Peru and Ecuador, producing a cold tongue of water that extends from South America all the way to Samoa.

With all that heat moving west, the rising moisture from the western Pacific creates such an intense low-pressure zone that it affects the monsoons off southeast Asia, drenches parts of Australia, and produces rains as far west as southern Africa.

The polar jet stream, which in an El Niño stays high in Canada, during a La Niña moves further south, driving frigid air deep into the United States. Freezing winters visit the Northwest and upper Midwest, less rain falls in the Gulf and southeastern states, and drought is common in the southwest desert.

During an El Niño, a tropical jet stream pushes across the Atlantic, shearing off the tops of hurricanes approaching the East Coast. But during a La Niña, hurricanes in the Atlantic encounter no westerly wind resistance and are twice as likely to strike the United States.

> **CAUTION**
>
> **Global Warnings**
>
> The 1998 La Niña hurricane season was the deadliest in the past two centuries. Hurricane Mitch killed 9,000 in Central America.

Keeping Track

The 1997–1998 El Niño marked the first time in human history that meteorologists were able to predict abnormal flooding and droughts months in advance, allowing time for threatened populations to prepare. The U.S. National Oceanic and Atmospheric Administration (NOAA) first announced the possibility of an El Niño as early as April 1997. Australia and Japan followed in May. By summer the details of those predictions were available to countries all over the world.

In Peru warnings reduced the number of tragedies. Grass grew on lands that were normally barren, so farmers could emphasize cattle. Rice and bean fields could be planted in areas normally too dry for those crops. Fishermen outfitted their boats for shrimp harvests, which the normally cool waters off Peru don't usually support. Even in some of the poorer regions of Peru, constructing storm drains, stockpiling food, and preparing for disease saved hundreds of lives.

Making those predictions involves a Herculean effort. One of the biggest helps has been the TAO (tropical atmosphere/ocean) array of 70 moored buoys which span the equatorial Pacific. Completed in 1994, the TOA buoys monitor water temperature from the surface down to 1,600 feet. The data collected by the buoys is transmitted to polar-orbiting satellites and then to NOAA's Pacific Marine Environmental Laboratory in Seattle.

Normally predictions are made on a statistical basis. Data collected from previous El Niños, for instance, may show that lower barometric pressure and higher sea-surface temperatures in Tahiti usually mean more rain for Ecuador and less rain for northern Brazil. By determining what conditions led to past El Niños, scientists can be on the lookout for similar match-ups in the future. But this kind of statistically driven approach doesn't tell scientists the mechanics involved.

With the advent of supercomputers, scientists are now able to input the data as well as fundamental laws of oceanic and atmospheric physics into simulated worlds and see how the weather changes over time. Scientists build these models based on present situations, inputting the data and seeing if the models can't spit back a world that scientists can observe. When the models are refined enough to do that, scientists put them to work predicting the future. With these models, scientists were able to predict the 1997–1998 El Niño almost a year in advance.

The Big Switch

Scientists believe that El Niños have grown more frequent and more intense during the last century. During the twentieth century there were about 23 El Niños and 15 La Niñas. The four strongest have occurred since 1980.

Is El Niño changing? The IPCC reports that the frequency and intensity of El Niño/La Niña events have increased since 1970 compared with the previous 100 years. The warm or El Niño phase of this team of weather phenomena has been relatively more frequent, persistent, and intense. Scientists believe this change began around 1976.

Global Warming

Has global warming been responsible for El Niño events? Scientists are uncertain if global warming thus far has had an effect on these events. But El Niño does produce more heat. That heat is transported out of the tropics during the peak of an El Niño and this raises global temperatures. With global warming there will be more heat available. The cycle might be shorter, because the recharge time is shorter. With world temperatures higher, the water off Ecuador and Peru will take less time to reach critical temperatures.

Whether the increase in the frequency and intensity of El Niño events is due to current global warming is uncertain. But if global warming increases in the twenty-first century, scientists predict there might be both more frequent and more intense El Niños to contend with.

The Least You Need to Know

- El Niño events are caused when the warmer water and lower pressure normally found in the western Pacific near Australia and Indonesia shift toward the eastern Pacific near Ecuador and Peru.

- Along the normally dry equatorial coast of South America, El Niño events ruin the offshore fishing and produce torrential onshore rains.

- El Niño events affect the weather worldwide.

- El Niños have increased in frequency and intensity since the mid-1970s.

- Scientists predict that if global warming increases, so will El Niños.

12

Atmospheric Variables

In This Chapter

- ◆ Other factors that might influence global warming
- ◆ Earth's erratic orbit
- ◆ Sunspots and droughts
- ◆ Hot times in the city

Despite all the careful calculations of well-meaning scientists, we have to be honest. The case for global warming, if we exclude the carbon dioxide evidence (a *big* exclusion), is based on global temperature rises of 1.1°F (.6°C) over the twentieth century. That's not a lot. Is it possible that the causes for this rise could be related to something else besides greenhouse gasses? A variety of astronomical, atmospheric, and geological effects have driven climate change over the history of the earth. In this chapter, I'll play the devil's advocate and look at some of the things that could be changing our weather right now, things that in fact *have* changed our weather in the past.

Continental Drift

When scientists discovered evidence of an ice age that occurred 250 million years ago in present-day Africa, Australia, South America, and India, they were baffled. These were equatorial countries. What was evidence of ice age glaciers doing on them? It wasn't until the theory of plate tectonics or continental drift came along that these ice age remnants began to make sense. According to the theory of plate tectonics, the earth's crust is made up of a number of small pieces or plates that are floating on a partially molten layer.

Pangea—the super continent.

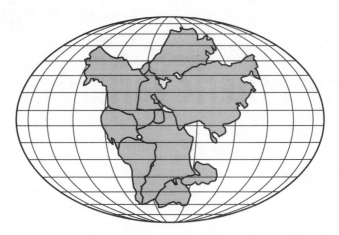

Scientists who formulated this theory believed that these continents were part of a giant super-continent called Pangea which began to break into pieces and drift apart about 200 million years ago. In fact, if you look at the landmasses, they seem to go together, like pieces in a puzzle. Pangea is probably not the only supercontinent that ever existed, but it's the one we know the most about.

Scientists believe that when this landmass broke up, the various pieces drifted apart on their own plates, eventually ending up where they are today. That explains how Africa, Australia, and India, which we find in subtropical latitudes today, could possess ice age evidence. They were once found much further south, near the Antarctic.

We now believe that continents drifting past or bumping up against each other are responsible for many of Earth's landmass forms, including its tall mountains, coastal bays, peninsulas, and islands. These drifting continents must have also had a drastic effect on ocean currents as well as the atmosphere, though in different ways, of course. Without the Himalayas and the Tibetan Plateau, the Indian monsoon might still be active, but it would be a lot less intense. In Chapter 6, I discussed how the movement of the landmasses may have contributed to Snowball Earth, a phase of Earth's development during which the planet was covered with ice.

But continental drift occurs at a snail's pace—a prehistoric snail at that. Continents might move two inches a year, often less. The effect on weather is measured in millions of years, not centuries.

View along the fault line.

(Photo by Michael Tennesen)

Asteroids and Dinosaurs

How about asteroids? Scientists believe asteroids have made big changes in the past. The theory that a big one impacted the earth about 66.4 million years ago and led to the extinction of the dinosaurs was first advanced by scientists who discovered a high concentration of the rare metal iridium in rock strata near Gubbio, Italy. When dated, the iridium-laced rocks coincided with the end of the dinosaur era. Further evidence of the iridium/dinosaur link was found in a number of sites located in Denmark and elsewhere. The only thing that could account for the iridium was material from an asteroid, which would have a much higher level of this mineral than the rocks found on Earth.

Hot Debates

Scientists believe that the giant Chicxulub crater in Mexico's Yucatan Peninsula could be the imprint of an asteroid that many believe initiated the extinction of the dinosaurs. Scientists, however, debate over whether this impact was sufficient by itself to wipe out the enormous reptiles, or if other factors were involved.

Scientists attributed this iridium concentration to a collision between Earth and a huge asteroid. An impact explosion with a large asteroid would have ejected an enormous volume of earth and asteroid into the atmosphere. This could have produced a cloud of dust and solid particles that would have encircled the earth and blocked out sunlight for years. The loss of sunlight may have killed off the plants that the plant-eating dinosaurs ate, which would have wiped out the predator and scavenger dinosaurs as well.

However, the asteroid theory doesn't seem to fully explain the dinosaur extinction. Some species of dinosaurs hung around quite a bit after the asteroid impact. It might be that to change the planet from Jurassic Park to what it is today required a number of changes—biological changes, climate changes, an asteroid impact, and maybe even a few volcanoes.

Asteroids can't explain the current level of global warming either, but they show you how climate can change quickly. Let's look at the earth's orbit for some possibilities.

The Earth Gets Around

The earth gets its heat from a fraction of the sun's heat. In fact, it receives less than two billionths of the sun's energy. The fact that the earth is moving in different ways affects how the sun's energy is received on Earth. The earth has basically two principal motions—rotation and revolution. It rotates as it spins on its axis, and it revolves around the sun. Both the rotation and the revolution affect climate.

The earth travels around the sun at approximately 70,200 mph (113,000 kph). Early man thought the earth was still, or otherwise we'd feel the air rushing past us. The reason we don't feel the air rushing past us is that the air, held on by gravity, is going just as fast as the earth.

The Seasons

The seasons are created by a change in the length of daylight hours. As the seasons progress, the days get shorter or longer and the noon sun changes its altitude in the sky. In summer the sun is high above the horizon as it makes its way across the sky. As winter approaches, the noon sun is lower in the sky and sunsets occur earlier each evening.

The seasonal variation in the angle of the sun affects the amount of solar radiation the earth receives. When the sun is overhead, its rays pass through a thickness of one atmosphere. But when the rays enter at an angle of 30 degrees, they pass through a thickness of two atmospheres, at 5 degrees they pass through the equivalent of

11 atmospheres. The longer the path, the more chance for scattering, absorption, and reflection, which rob the sun's rays of their intensity.

The annual changes in the sun's angle and the length of the day are created by the earth's angle of orientation toward the sun. The earth's axis is not perpendicular to the plane of its orbit around the sun. It is tilted 23½ degrees from the perpendicular. Remember your globes? They are tilted on an angle, and that angle approximates the earth's attitude toward the sun. The tilt is called the *inclination of the axis*. But that inclination wobbles a little. Right now the earth's axis is pointed at the North Star, Polaris. By about 14,000 C.E. it will be pointed at Vega, the new North Star. That's because of that wobble, and that wobble can affect the earth's orientation toward the sun and, as a result, its climate.

On the day of the *summer solstice* (June 21), the north pole is inclined 23½ degrees *toward* the sun. This is the beginning of summer in the northern hemisphere and the longest day of the year. In the southern hemisphere it's just the opposite—the beginning of winter and the shortest day in the year. Six months later on the day of the *winter solstice* (we're talking northern hemisphere again, folks), the north pole is inclined 23½ degrees *away* from the sun. This is the beginning of the winter, the shortest day of the year, the day upon which the people in the northern hemisphere receive the least light and energy from the sun. The summer and winter solstices occur on opposite days in the southern hemisphere.

> **Warm Words**
>
> The **inclination of the axis** of the earth is a measure of the number of degrees that the axis is tilted away from the plane of its orbit around the sun. In the northern hemisphere the **winter solstice** in on December 21, when the north pole is inclined 23½ degrees *away* from the sun. The **summer solstice** (again, we're talking northern hemisphere, folks) is on June 21, when the north pole is inclined 23½ degrees *toward* the sun.

Both the orbit of the earth around the sun and the tilt of the axis vary in certain ways. These variances may have caused some of the climate changes in the past. Could they be creating the observed changes in the present?

Eccentricity

The earth moves around the sun, but its orbit isn't exactly circular. Rather than a perfect circle, like a basketball, sometimes its orbit is out-of-round, like a football (not that much, but you get the idea). The out-of-roundness or *eccentricity* of the orbit changes the distance from the earth to the sun at different times of the year. Currently the earth is closest to the sun in January and furthest away in July.

However, we are in a period of low eccentricity—the earth's orbit around the sun is rounder—so the difference in the distance to the sun between January and July is slight (3 percent). That variance doesn't have much effect on the amount of the sun's energy the earth receives. But when the eccentricity of the orbit is high, the difference can be as much as 20 percent and that can have a much bigger effect on the atmosphere.

The change in the eccentricity of the orbit from a circular one to one more elongated takes about 100,000 years. What is interesting is that the ice ages also peak about once every 100,000 years—for the last million years or so.

Doing the Wobble

The earth rotates on its axis, but it's not a perfect rotation. The earth, like a spinning top winding down, wobbles a little. The wobble changes the direction toward which the earth's axis is pointed. Right now the north end axis of the earth is pointed toward Polaris, our North Star. However, in about 14,000 c.e. it will be pointed toward Vega, which will then be our new North Star. When the Egyptian pyramids were built, the earth's axis pointed toward Thuban. This wobble is called *precession*. One wobble takes about 19,000 to 23,000 years, and causes each end of the earth's axis to circumscribe a small circle in space.

Warm Words

Eccentricity refers to the shape of the earth's orbit, which varies from round to oblong. **Precession** refers to the amount of wobble in the earth's axis. **Obliquity** refers to changes in the tilt of the earth's axis.

Tilt!

Remember that seasonal tilt which causes the difference between summer and winter? I mentioned it was 23.5 degrees. Well, that's what it is now, but it actually varies between about 22.5 and 24.5 degrees. When the tilt is less, the seasons will vary less, and glaciers have a better chance of expanding. Shorter cooler summers will allow more snow to survive, and longer, milder winters would still be below freezing, but warmer air will produce more snow. The change in the tilt of the earth's axis is known as its *obliquity*.

The changes in the earth's obliquity, from low obliquity to high obliquity and back, take about 41,000 years.

Ice Age Evidence

Let's look at all this evidence. The earth's eccentricity or "out-of-roundness" increases and decreases every 100,000 years. In addition the earth's axis wobbles slowly, a process that happens every 19,000 to 23,000 years. And the tilt of the earth varies on a rotation that takes about 41,000 years. Although these orbital features don't affect the total sunshine received by the earth, they change where and during what season the sunshine is received.

So those changes in orbit and angle do affect the climate. But have they caused the amount of warming we've experienced in the last 150 years? Some scientists would have you think yes, they did. But most think not.

Climatoids

Scientists studying the bottom of the ocean are able to date the advances of the ice during the last ice age by measuring the chemical composition of the shells in each sediment layer. What they find is that over the last million years, ice grew for about 90,000 years, shrank for about 10,000 years, and repeated, with smaller disturbances about 19,000, 23,000, and 43,000 years apart. These statistics correspond to the orbit and tilt of the earth around the sun.

Solar Cycles

Then there's the sun. Its output is variable. The problem is we weren't able to determine the variability of the sun's output until we were able to watch it from outside our atmosphere. We're doing that now from satellites, but we haven't been at it long enough. The only solar patterns we've looked at over time are sunspots.

Sunspots

Sunspots are some of the best studied of any of the features of the sun. They appear as dark areas simply because they are somewhat cooler than the surrounding gases. These spots are related to magnetic activity in the sun's outer layers, but the processes are not fully understood. Sunspots can appear in groups of two to more than 100. The average lifetime for a given sunspot is about one week. Sunspot cycles occur about every 11 years on average. When sunspot activity is high, the regions surrounding sunspots can have a number of solar phenomena including solar flares (bursts of high-energy particles) and prominences (huge solar clouds that erupt from the surface). (I talked about them in Chapter 3.) During times of active sunspot cycles, the earth gets a bigger dose of solar energy.

Climatoids

The low sunspot period from 1645 to 1715 corresponds to a very cold interval over the entire globe. It was the coldest period of the Little Ice Age. British astronomer E.W. Maunder discovered it. Today it is referred to as the Maunder minimum.

People have been looking at sunspots for 2,000 years. Galileo paid particular attention to these little dark spots on the sun. We know that sunspots increase and decrease on a regular cycle, about once every 11 years. But in the historical records kept by scientists, there have been some periods with few or no sunspots. One of those periods was from 1645 to 1715. It turns out that this was also the coldest period of the Little Ice Age. It is referred to as the Maunder minimum.

Could this be the magic bullet? Some scientists point to the fact that the second half of the twentieth century has been a period of relatively high sunspot activity.

Droughts and Tree Rings

In addition to the 11-year sunspot cycle, there is also another 22-year cycle in which the magnetic polarities of sunspot clusters reverse. Scientists who study the tree rings of the bristle cone pines in the White Mountains of California believe they have uncovered a sequence of droughts in the rings that match the magnetic polarity reversals in the sunspots. The idea is controversial. Scientists can determine the amount of precipitation in each year by the width of the ring in the core of the tree. Wider rings represent happier trees, which in the arid west means periods of higher rain. It's not always that simple. "Happier" means different things to different trees in different areas. Furthermore, observations of other phenomena don't corroborate the magnetic polarity/drought theory.

Volcanic Ash

Scientists believe that volcanic eruptions have the effect of changing the climate of the earth. Big, explosive eruptions can send huge quantities of gas and fine debris into the atmosphere. The bigger ones have sent debris high into the stratosphere where it spreads around the globe and might remain for months or even years. The debris has the ability to filter out a portion of the sun's energy and to lower the earth's temperature.

The most notable cold spell linked to a volcano was the one that followed the eruption of Mount Tambora in Indonesia in 1815. Many scientists attribute the abnormal cold spring and summer of 1816 to the clouds of volcanic debris put into the atmosphere by Mount Tambora. There were other less dramatic effects from Krakatoa in 1883 and Mount Agung in 1963. However, it wasn't until the eruptions of Mount St. Helens in the state of Washington in 1980 and El Chichòn in Mexico in 1982 that scientists really got to study the volcanic blasts with the use of sophisticated satellites and remote sensing instruments.

Mount St. Helens

Prior to 1980, Mount St. Helens, Washington, was a steep conical volcanic peak that rose 9,680 feet (2,950 meters), had a snow-capped summit, and even a few small glaciers. But on the morning of May 18, 1980, the entire north side of the summit came down—about 0.5 cubic mile (2 cubic kilometers) of rock and ice. An instant later, an enormous explosion of expanding steam and volcanic gas rocked the countryside. The gases formed a ground-hugging black cloud filled with hot, dense debris that raced over four major ridges and valleys up to 17 miles (28 kilometers) from the volcanic summit.

I visited the site after the eruption and the devastation was awesome. For the first couple miles from the summit, everything had been obliterated. All you could see was a blanket of ash. A horseshoe-shaped crater 1.2 miles across (2 kilometers) and roughly 2460 feet deep (750 meters) replaced the peak. But the most impressive thing was the blowdown zone where huge virgin Douglas firs were snapped like matchsticks and lay on their sides, covered with ash. The U.S. Forest Service estimated that 10 million trees were felled by the blast.

CAUTION

Global Warnings

Though the destruction caused by the eruption of Mount St. Helens was enormous, the total magma displaced was only about .48 cubic mile (two cubic kilometers). The explosive eruption of Tambora Volcano in Indonesia in 1815 displaced about 12 to 24 cubic miles (50 to 100 cubic kilometers) of magma. That bested the 1883 eruption of Krakatoa, which displaced about 4.5 cubic miles (18 cubic kilometers). But civilization has never been tested by a cataclysm on the scale of the enormous eruption at Yellowstone about 2 million years ago when over 720 cubic miles (3,000 cubic kilometers) of boiling magma exploded into the air. Back further, they're even more enormous.

When Mount St. Helens erupted, there was instant speculation that it could have major effects on the climate. For a time the large volume of volcanic ash that it emitted had significant effects both locally and regionally. But worldwide cooling was less than 0.2°F (0.1°C).

El Chichòn

However, studies following the eruption of El Chichòn showed a worldwide cooling effect on the order of .5 to .9°F (0.3 to 0.5°C). Why, if it was less explosive than Mt. St. Helens, did it have a greater impact on global temperatures? The reason is that the material emitted by Mount St. Helens was largely fine ash that settled out relatively quickly. El Chichòn, on the other hand, spewed an estimated 40 times more sulfur-rich gases than Mount St. Helens. These clouds combined with moisture in the stratosphere to produce dense clouds of sulfur acid droplets, which both absorbed and reflected solar radiation. Explosiveness alone is a poor instigator of climate change. For volcanism to impact the climate, you would have to have a whole bunch of volcanoes going off over a relatively short period of time. Such an idea was once floated as the reason for the ice ages, but scientists have other ideas today.

Water and Carbon Dioxide

Everybody is looking at carbon dioxide as the principle villain for the amount of global warming we've seen thus far. But if an increase in greenhouse gases causes global warming, then the warming will create more evaporation. Water vapor is a powerful heat absorber, so the additional water vapor could compound the problem—more heat leading to more evaporation leading to more heat. It could produce a runaway effect. Or the increased water vapor could lead to more clouds, which could block incoming radiation. And that could serve as a check on runaway radiation. Scientists haven't figured this one out yet.

Urban Heat Island Effect

Maybe it's us. Maybe we are the ones causing the extra heat. At the beginning of the last century, only about 2 percent of the population lived in cities of more than 100,000 people. Today it's more like 25 percent worldwide. In places like the United States, western Europe, and Japan it's better than 50 percent. The *urban heat island effect* is one of the most studied and documented climatic effects on Earth. What this means is that temperatures within cities are generally higher than in outlying rural areas.

> **Warm Words**
>
> The urban heat island effect is a term used to describe the fact that urban environments are usually warmer than surrounding rural areas.

The effect is more pronounced at night, with minimums in the cities averaging 0.6 to 0.9°F (1.2 to 1.7°C) higher. But that's an average. On calm nights the temperature difference between the city center and the countryside in Washington, D.C., can be as much as 20°F (11°C). The tall buildings, concrete, and asphalt of the city absorb and store greater quantities of solar radiation than do the vegetation and soil typical of rural areas. At night the hard surfaces of the city release their heat more slowly than the earth. Heating, air conditioning, power generation, and transportation release a lot of waste heat. One study of a built-up portion of Manhattan found that during the winter the quantity of waste heat was 2.5 times the amount of solar energy reaching the ground.

The urban heat island effect.

(Photo by Michael Tennesen)

In addition, industry, cars, and heating appliances release a blanket of pollutants over the city, including particulates, water vapor, and carbon dioxide. These greenhouse gases make the city warmer than the surrounding territory. They also act as condensation nuclei, which encourage precipitation. Not only is it hotter in the city, it rains more.

What effect does the urban heat island have on temperatures worldwide? It doesn't affect the world as a whole, but most temperature readings over the last 100 years have been taken in cities. The increase in temperature could simply be the effect of the cities on the temperature-takers. Though the IPCC says it has compensated for the urban heat island effect in its readings, they also admit that determining accurate temperatures from past readings are problematical. They involve temperatures taken with different instruments at different times and with different standards. Add to that the urban heat island effect and you have the makings of confusion. Could the total of all these variations, including continental drift, the earth's orbit, solar variation, volcanic ash, and clouds be responsible for the 1.1°F (.6°C) change scientists have recorded in the past century? Maybe. Maybe not.

The Least You Need to Know

- The drifting of the continents on the surface of the earth and impacts by asteroids have altered the climate in the past.

- The out-of-roundness of the earth's orbit, the changing tilt of its axis, and the wobble in its spin have altered the advances and retreats of ice age glaciers.

- Solar output, sunspots, and changes in magnetic polarity can affect our climate.

- Volcanic emissions can put a chill in your evening.

- Urban environments are hotter than rural areas and this may have had an effect on past temperature readings.

- All of these aspects might have played a part in the amount of global warming we've experienced to date.

Ratcheting Up the Equation

In This Chapter

- Things that could make it a lot worse

- The critical role of ocean currents

- Shutting down the currents

- A catastrophe in the making

In the previous chapter, I gave you some alternative explanations for the global warming that has occurred in the last century. So maybe things are not so dire. Maybe the scientists have it wrong. Maybe you can relax. Toss some fossil fuels in the fireplace. Take it easy. But before you slide into your easy chair, I want you to be aware of some natural triggers, which just in case this global warming thing is for real, could make it a lot worse. Maybe you'll want to sit up a little.

Scientists who've studied climate change over the ages tell us that our climate is not a rheostat. It's not one of those dimmer switches that you turn down slowly. It's more like a flip switch. As you apply ever-increasing amounts of pressure to the switch, it hits a point where it flips, the lights go off, it's dark, and you are in another state altogether. In this chapter, I'll look at some of the feedback mechanisms in the atmosphere that could ratchet up the pressure on that light switch and hasten the flip to another state. Then I'll look at ocean currents, particularly the ones in the deep north Atlantic, investigate what things could cause a change, what changes have happened in the past, and what we're liable to encounter in the future.

Feedback Mechanisms

We live in a feedback-dominated climate system in which living things, land surface, ice, oceans, and the atmosphere are all linked up and interrelated. Within this complex system are various mechanisms, which balance each other out. The body is a good example of a regulated feedback mechanism. If we exercise, we get hot, but then perspiration forms, which tends to cool us down. If you stopped sweating while exercising in warm temperatures, you could get heat stroke. We need the feedback mechanism that perspiration provides to control the buildup of heat. When feedback mechanisms are turned off, this can unleash runaway processes. Snowball Earth, which I discussed in Chapter 6, was an example of a feedback mechanism—the buildup of ice to limit the removal of carbon dioxide from the atmosphere—which when turned off, led to catastrophic results for all life on planet Earth.

If you recall, during Snowball Earth, the continents had drifted northward from their location nearer the Antarctic to positions mostly scattered around the equator where it rained a lot. And the rain removed the carbon dioxide from the atmosphere, which reduced the greenhouse effect, which caused the planet to get colder. Ice packs formed in the polar oceans, and these reflected sunlight, driving temperatures even lower. A feedback process got started that proved unstoppable and soon the planet was engulfed in ice.

> **Warm Words**
>
> **Carbon burial** is the process whereby silicate rocks erode and the chemical breakdown of the rocks converts carbon dioxide to bicarbonate, which is then washed to the oceans where it ends up as carbonate sediments at the bottom of the sea.

The unusual configuration of the continents near the equator may have been the reason that Snowball Earth got going in the first place. When the continents are closer to the poles, and the ice is advancing, the ice covers the land and stops the carbon from being washed into the sea. This keeps the carbon dioxide high in the atmosphere, which does its greenhouse job of warming up the place enough to prevent a runaway freeze. But with the continents down near the equator, ice doesn't cover them, the *carbon burial* process keeps lowering the carbon dioxide level, and the critical threshold for a runaway freeze is reached because the carbon dioxide "safety switch" fails when carbon burial continues unchecked.

Albedo

How well sunshine is reflected back from a given surface is what we call the "albedo" of that surface. If you had 0 percent albedo, that means none of the sunshine is reflected back, the surface absorbs it all, and it gets hotter. However, 100 percent albedo means that all the sunshine is reflected back, none of the heat is absorbed, and the surface stays cold. Black surfaces have 0 percent albedo; white surfaces have 100 percent or thereabouts.

Clouds reflect a lot of sunlight and therefore have a high albedo, particularly low, thick ones. The altitude and type of the clouds affects albedo. If we were to reduce the amount of clouds on Earth, we would reduce its albedo, the earth would suck up all the heat, and it would get warmer. Dust and particulates influence clouds, and an increase in these elements can make it easier for clouds to form, which might reflect the sun's energy backward. The increase of sulfates in the air after WWII increased the clouds, which some scientists feel might have caused a mid-century cooling in an otherwise warming century. Snow and ice also have a high albedo and thus reflect a lot of sunlight. The accumulation of snow and ice during Snowball Earth reflected sunlight, which led to the formation of more snow and ice.

> **Climatoids**
>
> Snow and ice reflect heat but the rainforest absorbs it. Thus, the removal of the rainforests might have the short-term effect of decreasing the heat absorbed at the surface, but the release of carbon dioxide from cutting down or burning the timber will have a net effect of more warming.

The Bogs Are Brewing

English scientists report that peat bogs across the northern latitudes of Europe, Siberia, and North America may be decomposing and releasing carbon dioxide into the atmosphere. Scientists believe that as the bogs dry out, they might trigger an enzyme, which promotes the decomposition of organic carbon. Peat bogs in the upper latitudes hold some 450 billion tons of carbon, which is the equivalent of 70 years of industrial emissions.

Scientists have recorded a 65 percent increase in the release of carbon from British peat bogs in the past 12 years. The carbon locked up in

> **Global Warnings**
>
> Scientists worry about the amount of methane, one of the greenhouse gases, which is locked up in the permafrost (permanently frozen ground) in the vast stretches of Arctic tundra spread out over the high latitudes of North America, Europe, and Asia. If the permafrost melts, the resultant release of methane could accelerate global warming even further.

soils for thousands of years is starting to escape and this escape may hasten global warming. Scientists believe much of it will end up as carbon dioxide in the atmosphere, which will only increase the greenhouse effect, the warming, and further accelerate the decomposition of the bogs.

Ice Melt Kills

Researchers agree that greenhouse warming will cause a substantial rise in sea level due to the acceleration of the ice melt on the land and at the poles. Some scientists, however, contend that the increase in moisture caused by the warming will result in an increase in snowfall over Antarctica, which could counterbalance melt water by removing water from the sea to supply the snows.

CAUTION

Global Warnings

If all the ice in the world were to melt, it would result in a rise in sea level equal to 230 feet (70 meters), enough to drown most of the major ports of the world.

But the worst case scenario, according to some scientists, is the breakup of the west Antarctic ice sheet. This mass of ice is a mile thick and spreads over an area the size of India. It rests near the sea floor like an enormous grounded ship. As the climate becomes warmer, the west Antarctic ice sheet could break up and possibly melt. It would take a while, but once started there'd likely be no stopping it. The result could be a rise in sea level of 10 feet (3.4 meters) worldwide.

Ocean Currents

As I mentioned in Chapter 10, the earth's climate is forever trying to rectify the heat imbalance between the tropics and the poles by shuffling heat from the equator to the polar ice caps. We talked about the part that the atmosphere plays in this heat exchange, but the atmosphere is only responsible for about half the heat. The other half is transported through the oceans. The water around the equator is heated more than any other ocean water, and the ocean moves that heat toward the poles to even out the heat budget.

Most of the surface waters of the ocean are moved by friction created by the winds. They push on the surface of the ocean and eventually produce large currents that move water clear across the ocean. The trade winds push the Pacific waters from Peru to Australia. They also move Atlantic waters from Africa across to South America. The Gulf Stream of the Atlantic and the Kuroshio Current of the Pacific are some of the major currents created by wind.

Not only does heat and circulation play a role, but salt is a factor as well. Salty water is denser or heavier than normal water, and this is a factor in the great conveyor belt–like circulation of the deep oceans. If you place dense water over less dense water, the denser water will sink, creating a current even without the help of the wind. You can make water denser or heavier by making it colder or saltier. Water becomes saltier when evaporation removes the moisture but leaves the salt behind.

The Big Conveyor Belt

Surface currents aren't the only things going on in the sea. There are also global ocean currents that circulate on the surface and in the deep sea, connecting the major oceans on the planet. The scientific name for this conveyor belt–like circulation is the *Thermohaline Circulation (THC)*. In this current, warm salty Atlantic water near the equator moves north toward Greenland and Labrador where it cools and sinks. The sinking occurs in a couple of relatively small places in the sea, diving more than a mile deep before the water flips around and heads south back down the Atlantic.

The water flows south and swings around the tip of Africa, rising to the surface in the Indian and Pacific Oceans, as well as in certain spots around the ice shelf off of Antarctica before the water heads back toward the Atlantic and north again.

Warm Words

The **Thermohaline Circulation (THC)** is a global ocean current that operates like a conveyor belt running through shallow and deep-sea waters to connect the major oceans of the planet.

The Critical Sinks

The Atlantic waters in this system that begin at the equator and travel north arrive at two relatively small spots, one east of Greenland and one east of Labrador. At this point the water has grown heavy and dense enough from the cool air and the salt it brings with it to sink to the bottom.

The only other place where the waters descend along this conveyor belt is off the ice shelf around Antarctica. There the freezing of the water along the shelf takes up water but not salt, leaving a bath of salty water for the deep waters that rise up to the surface here. Those waters linger in the salt, lose a little heat, and make another dive again, continuing their long journey again. There are no such sinks in the Pacific Ocean because the water is simply not salty enough—a prerequisite if the water is going to sink to the bottom.

The greatest effects on the atmosphere by this global ocean conveyor belt, the THC, are made in the north Atlantic where the heat from the tropics is transferred north and released into the atmosphere before it sinks. This is much greater than the heat exchange that occurs in Antarctica where already cool water rises to the surface and cools a little more.

Jamming the Works

So how do you stop the deep ocean conveyor belt, the THC? Richard Alley in his book, *Two-Mile Time Machine*, compares the deep ocean conveyor belt to a grocery store conveyor belt and asks what would you do to stop *that* conveyor belt. His suggestion is to grab a fork and stick it into the small gap where the grocery store conveyor belt goes down. The conveyor belt will pull the fork down with it and get stuck.

Salt, or the absence of it, works like a fork to clog up the conveyor belt at the point in the north Atlantic where the water sinks to the bottom. The flow of water from the tropics to the north Atlantic exists within a delicate salt balance. The tropical ocean sends warm salty water to the north Atlantic, but at the same time it sends water vapor north which falls out as rain and snow. That moisture falls on the ocean or flows to it from land and decreases the Atlantic's salinity. Remember, the Atlantic surface waters are saltier than the Pacific. Also the patterns of rain and runoff are such in the North Atlantic that they create this problem. It's a tricky situation. If enough fresh water is added to the mix, then when the cooling waters hit the north they will be cool enough but not salty enough to displace the water beneath it.

Remember, water needs to be cool and salty to become dense or heavy enough so that when it arrives in the north Atlantic, it sinks to the bottom and continues the motion of the conveyor belt, the THC. As the water sinks to the bottom, it draws up more water from the equator bringing its gifts of heat and moisture.

So how's global warming going to affect the THC? It will not only increase the warmth of the water, but, even more importantly, it will increase the amount of precipitation. This fresh water is dangerous since it will in general decrease the percentage of salt in the north Atlantic by diluting it with fresh water.

Then what? Well, if the water gets fresh enough, the sinking will slow or stop and water that arrives from the tropics, cooled by the atmosphere, will simply pool and possibly freeze. The ice will increase the albedo, and it will get even colder. Global warming will not be warm for everyone.

What If the Conveyor Belt Stops?

If the THC were to stop, it would also stop the transfer of heat from the tropics to the northern latitudes of the North Atlantic. Some scientists suggest that the heat given off by this current is the reason that northern Europe is at the same latitude as Alaska, yet is much warmer. This current influences the eastern or European side of the Atlantic more than the western or New England side.

There are, however, other influences. The prevailing westerly winds blow over the ocean and spread its warmth over the west coasts of the continents. The jet stream also brings Arctic air down over New England and the Asian coast of the Pacific but brings warmer breezes to the U.S. Pacific Northwest and Europe.

All that said and done, the THC is an important influence in the climate of the North Atlantic. Complete shutdown of the THC would remove an enormous amount of heat that is transferred from the tropics to the North Atlantic. If the THC were shut down, it could create a net loss to the region as high as 7.2°F (4°C). This could increase the amount of sea ice, and that could lead to even more cooling. (Remember, ice reflects heat.)

> **CAUTION**
>
> **Global Warnings**
>
> The shutoff of the global ocean conveyor belt, the THC, in the North Atlantic would lower the temperature of the region by about 7.2°F (4°C). That's the difference between this age and the ice age. How much is that? Global warming might signal hot times for the rest of the world, but a deep freeze for the North Atlantic.

Dead Roses for England

Great Britain is downwind of where the conveyor belt dumps its gift of warmth and moisture. Subsequently, British citizens grow roses at the same latitude where polar bears live near the Hudson Bay. The warm, salty water that enters the North Atlantic must get rid of its heat to make the dive, and England and much of northern Europe gets to share in that bounty.

> **Climatoids**
>
> Wind blowing over warmer waters and carrying heat and increased moisture is familiar not only to people who live in Great Britain, but to residents of Buffalo, New York, and Erie, Pennsylvania. In these places, subzero winds that blow down from the Arctic in the winter encounter the warm waters of the Great Lakes and pick up moisture and heat. They then dump their gift on communities below the lakes. The result is that while people out on the Great Plains sometimes suffer –40° weather, people in the snow belt under the Great Lakes are covered with a blanket of white that is just a little below freezing.

In models of the ocean and atmosphere, if you add enough fresh water to the mix of the conveyor belt that feeds the North Atlantic, the water there sinks more slowly. If the sinking slows enough and precipitation continues, the sinking could stop. Once the conveyor belt is shut off, its transport of heat to the north is terminated. The cool water pools on the surface and may freeze in the winter. This reflects heat and makes it even cooler. So once it's shut off, feedback mechanisms are engaged, which make it really difficult for the circulation to get going again.

According to Lynn Talley, Professor and Research Oceanographer at the Scripps Institution of Oceanography, University of California, San Diego, the results of a shutdown would be catastrophic. Eventually enough ice would freeze on the land to remove the fresh water that caused the dilution, but it would take about 1,000 years. That means that the Royal National Rose Society in Great Britain could be on hiatus for an entire millennium.

It would also decrease the rising of summer-heated air from Africa, which flows into Asia and drives their summer monsoons. A cold north and a drier Africa and Asia would decrease the wetlands, which would decrease the production of methane, which, as you recall, is another greenhouse gas. Reduction of greenhouse gases makes it even colder. Colder air typically carries less water, so that much of northern Europe could be subject to both drought and bitter cold.

Permanent El Niño

I know what you're saying. Wait a minute. You thought we were talking about global warming. How do we get from global warming to ice age Europe? Well, I just told you. What you really want to know is where would all that heat go? Well, like ducks flying south for the winter, the millennial winter that could result from shutting down the global ocean conveyor belt, the THC, would drive the heat south across the border.

Parts of the global ocean conveyor belt have shut down a number of times going back a million years or more. Not from global warming, but from other factors that upset the delicate warm, salty water mix. When it has shut off, parts of the northern hemisphere have gotten colder while parts of the southern hemisphere got warmer, even during the ice ages.

So what could that heat do? All the things that extra heat does. It would increase precipitation along the equator, decrease precipitation in drier inland areas. If the north decreases its temperature by 7.2°F (4°C), then the world would seek equilibrium and the south could increase its temperature that much on average. The South Atlantic, which would have no place to send its warmth, might sit below the equator and get hotter. This could have momentous effects on tropical storms, including hurricanes.

If we take 7.2°F (4°C) from the north and give it to the south, that amount of increase might be sufficient to kick in an El Niño. Only this El Niño could be permanent.

The Younger Dryas Event

If this stuff sounds all too fantastic or hard to believe, consider that just such a fresh water dilution occurred just prior to the Younger Dryas event. Though the cause was different, the mechanics were similar. The ice that advanced down the North American continent during the last ice age had the effect of blocking the rivers that flowed north and east into the Atlantic. The rivers that flow from the Great Lakes were so clogged that the lakes rose to a point where they overflowed into the Mississippi River and eventually ended up in the Gulf of Mexico.

But when the ice began to melt, the eastern outlets down the Susquehanna, the Hudson, and finally the St. Lawrence Rivers gradually opened. When water breaks through an ice dam, the flow of the water past the ice creates frictional heat that rapidly opens an ever-larger hole in the ice. Ice dams are recipes for catastrophe, and this was no exception. A huge flood raced down the St. Lawrence into the sea as the Great Lakes broke through the ice and sought out a new level. This enormous dumping of fresh water into the North Atlantic would have reduced the salinity of the poleward-moving currents and is a likely suspect for turning off the conveyor belt. A huge flood did occur down the St. Lawrence River just before the Younger Dryas, and scientists believe they are related.

Remember Younger Dryas from Chapter 6? Ice core evidence from Greenland shows that wind-blown dust and sea salt increased by a factor of 3 to 7 times. In Norway, mean July temperatures were about 12–16°F (7–9°C) colder than today. In Spain, the July temperatures were as much as 14°F (8°C) colder than today.

Younger Dryas affected North America as well. It threw us back into the ice ages for over 1,000 years. The greatest catastrophe of the Younger Dryas event may have been the extinction of the great ice age mammals. Though early human hunters definitely played their part by over-hunting, the mastodons, mammoths, horses, and saber-toothed tigers, which once wandered the North American continent became extinct during the Younger Dryas.

The Gamble

There might be switches that have catastrophic effects as extreme as the global ocean conveyor belt, the THC. It may just be *one* of the major climate switches that scientists have uncovered in their search to understand climate change in the past. According to scientists studying this phenomenon, flipping the THC on and off doesn't seem to account for all the drastic climate changes the earth has seen in the past. There must be other switches, other dangers lurking out there that we don't know about … yet.

We are engaged in a radical experiment, a lethal gamble. We are on the way toward doubling the amount of carbon dioxide in the atmosphere, and scientists who study the atmospheric past tell us that such radical increases in this and other greenhouse gases in the atmosphere in the past have had wide and catastrophic results. Though the future looks grim, it may be that the global warming we've experienced won't continue and the scientists are wrong.

It's a lot like staring down the barrel of Dirty Harry's 44 magnum wondering if it's really loaded and he's directing you to ask yourself one question. "Do I feel lucky today? Do you, punk?"

The Least You Need to Know

- Various triggers in the atmosphere could act as feedback mechanisms to increase or decrease the causes and effects of global warming.

- If the west Antarctic ice sheet broke up and melted, it could raise global sea levels by 10 feet (3.4 meters).

- A current that operates like a huge conveyor belt connects most of the oceans of the planet.

- Global warming could decrease the salinity in the North Atlantic, which could turn off the global ocean conveyor belt.

- Scientists believe that the global ocean conveyor belt stopped just prior to the Younger Dryas event, which threw the world back into an ice age for over 1000 years.

- If global warming were to switch off the deep ocean conveyor belt again, it would be a catastrophe.

Part 4

The Crystal Ball

So I've looked into the past, taught you a little about our weather and climate, and now I'm going to look into the future. I'll show you what the Intergovernmental Panel on Climate Change, a project of the World Meteorological Organization and the United Nations Environment Program, predicts global warming may do to Africa, Asia, Europe, and Latin America. How will it affect forests, agriculture, and tourism? Will it sink low-lying island nations?

Then you'll see what the U.S. Environmental Protection Agency predicts will happen in North America. What effects will it have on ranching, fisheries, and human health? Will Florida sink beneath the sea? I'll take a look at what the National Wildlife Federation predicts might happen to our wildlife, plants, and national parks.

I'll also investigate climate models and how they are used to foretell the future. If scientists can't predict the weather more than two weeks out, how do they propose to predict the climate over the next 100 years?

14

The Global Picture

In This Chapter

◆ Tough times for Africa

◆ Asia gets drenched

◆ Latin America gets sick

◆ The islands go down with the ship

I've discussed the worldwide effects of global warming, but what do scientists think global warming will do to specific regions? Human populations and natural systems differ radically in their vulnerability. An increase in rain will cause an increase in the number of mosquitoes and the number of mosquitoes carrying disease, but each region will be susceptible to different diseases and even different mosquitoes.

Regional differences will depend upon conditions and locations. Do you live in a desert? A swamp? A ski resort? A tropical island? All of these different areas will be affected by current weather, the change that global warming will make to it, and the resources at their disposal. A ski resort may be able to live with a little less snow, but a tropical island won't survive a 3 ft (91 cm) sea level rise if it's only 2 ft (60 cm) above sea level in the first place. Some areas will get wetter. Elsewhere, a desert may rise. In this chapter, you'll see what could happen to different regions around the world if the climate changes in the next century as much as some scientists predict.

Africa

Africa is a diverse region with different climates, geography, culture, and economies. It is predominantly tropical, hot, and dry. Tropical rainforests exist in the west and central portions along the equator. But as you go north or south, the land gets more arid. Most of the population lives in the sub-humid areas, north and south of the tropical forest, and in the semi-arid areas a little north and south of that. Different regions have different problems.

Deserts may rise.

(Photo by Michael Tennesen)

Tough Times for Agriculture

More than half of the people of Africa live in rural communities and are dependent on local crops harvested from nearby fields. Irrigation is inadequate, and many farmers have to rely on natural rainfall. While the rest of the world keeps figuring ways to grow more food, Africa's ability to feed itself keeps declining. Over the past two decades, Africa has suffered widespread famine and malnutrition. Its nations frequently have to rely on emergency food.

In the drought years of the mid-1980s, African nations ate about 50 percent more than they could grow. In the mid-1990s consumption exceeded domestic production by 30 percent. A third of all Africans eat less than 2,000 calories a day, which most residents of industrialized nations would call a diet. The continent needs to increase its agricultural output by 4 percent a year just to meet basic needs. But to achieve that, it may have to double its farm lands by 2050. And that puts more pressure on the tropical rainforests.

Now add some global warming and what have you got? Frequent and prolonged droughts would devastate agriculture, promote famine, and place the population at peril. There has been chronic war in several regions of Africa for the past three decades. Africa simply can't cope with the added stress. Government agencies are underfunded and undermanned. In many instances military coups, despotism, tribalism, corruption, and the stringent economic constraints of the international financial community have brought them to the brink. Africa's troubles would only be exacerbated by the additional burdens of global warming.

> **CAUTION**
>
> ### Global Warnings
>
> Food aid from outside countries constitutes a major portion of trade in Africa. In the 1990s, food aid constituted two thirds of the food imports for both Kenya and Tanzania. How will global warming affect this recurring crisis?

More Disease

Africa is susceptible to the climate changes created by El Niño. If the world's climate were to shift into a permanent El Niño, as some scientists predict, Africa would suffer. Following the 1997–1998 El Niño, malaria and cholera outbreaks were recorded in many countries in East Africa, the portion of the continent most affected by an El Niño event. In years of heavy rain in Rwanda, the incidence of malaria increased by 337 percent, and flooding in arid areas could promote the growth of malaria-bearing mosquitoes. In the Sahel region, which has suffered from drought over the past 30 years, malaria has decreased as still-water breeding sites for mosquitoes have dried up. But bring in a flooding, let the water pool up, and malaria could make a comeback.

Extreme weather brings disease. Floods contaminate public water supplies, and drought encourages unhygienic practices due to a shortage of water. Cholera was epidemic during the 1997–1998 El Niño. Africa then reported 80 percent of the total number of cholera cases worldwide. Meningitis, on the other hand, is associated with warming and reduced precipitation. So while the flooded parts of Africa may be suffering from cholera, the drought-stricken parts could come down with meningitis.

CAUTION **Global Warnings** _____

Following the 1997–1998 El Niño, Rift Valley fever, a disease of domestic animals, broke out in East Africa and killed as much as 80 percent of the livestock. The disease was thought only to affect livestock, and sometimes domestic animals, but never people. But the El Niño Rift Valley fever outbreak spread to livestock owners and rangers. What other surprises await if climate changes set in on a more permanent basis?

Asia

Asia would have problems with global warming. It already has more than 60 percent of the world's population, its natural resources are stressed, and its ability to adapt is poor. Many Asian countries are dependent upon natural resources like water, forests, grasslands, and fisheries. Sulfate emissions from Asia's rising industrialization could dampen some of the effects of global warming by filling the atmosphere with tiny droplets of sulfuric acid that will reflect away excess sunlight. But the acid rain that sulfates produce eats away marble and is deadly to inland fisheries. Using one toxic substance to counter another doesn't make a lot of sense.

The Sea Rises

More than half of the continent's population, 1.7 billion people, lives in the coastal areas of Asia. In countries like Bangladesh, Nepal, the Philippines, Indonesia, and Vietnam, many river watersheds suffer badly from deforestation, indiscriminate land conversion, and soil erosion. Global warming would impact these ecosystems by accelerating sea level rise and more frequent and more intense storms.

Low-lying coastal cities like Shanghai, Jakarta, Tokyo, Manila, and Bangkok would also be affected by sea-level rise with its saltwater intrusion, increases in silt, and the loss of land. Tropical Asia already suffers from some of the worst effects of El Niño, which produces forest fires and drought. There are at least 10 major deltas in the coastal zones of Asia with an area of more than 3,844 square miles (10,000 square kilometers) each. Low level deltas are especially sensitive to rising seas and pounding surf. With resources stressed to the max, little ability to adapt, and weak economic structures, global warming could wreck tropical Asia.

CAUTION **Global Warnings** _____

Bangladesh could lose close to 30,000 square kilometers of its delta regions based on projected sea-level rises of 39 inches (100 centimeters), while Indonesia may lose 11,550 square miles (34,000 square kilometers) and Vietnam 15,400 square miles (40,000 square kilometers). Such a sea-level rise would expose 13 percent of Bangladesh's population, 2 percent of Indonesia's, and 23 percent of Vietnam's. That's a total of 34 million people for all three countries.

Tough Times for Agriculture and Aquaculture

In general, things may get better for mid- and high-latitude regions where lengthening of the growing season and an increase in carbon dioxide could potentially help production of some crops. But for lower latitudes exposed to El Niño and tropical storms, it will get worse. Yields of several major Chinese crops are expected to decline. Acute water shortages combined with increased heat will be rough on wheat and rice production in India, even with the beneficial effect of more carbon dioxide. Crop diseases such as wheat scab, rice blast, and sheath and culm blight could spread in temperate and tropical Asia as the climate becomes warmer and wetter.

Asia produces 80 percent of the farmed fish, shrimp, and shellfish in the world. Wild stocks are over-fished. Trawlers destroy the sea bottom. And coastal development and pollution further stress the habitat. El Niño events already affect plankton, sardines, and prey species that are vital to marine ecosystems. Conservation of marine resources is behind schedule, and global warming will only make that worse.

Australia and New Zealand

Australia and New Zealand cover a region that spans the tropics and the mid-latitudes of the southern hemisphere. The region has a variety of climates and ecosystems including rainforests, deserts, alpine areas, and coral reefs. The surrounding oceans strongly influence both countries. Australia already has significant desert areas, and periodic droughts plague agriculture. New Zealand is smaller, more mountainous, more temperate, and might prove more resilient to global warming.

Increased carbon dioxide may initially boost agricultural output, but that could change as droughts and floods get more severe. Plus climate change may help spread pests and diseases. If droughts increase, it will hit rural communities the hardest. Droughts will increase the chance of forest fires. Basically, the initial advantage of a little global warming and an increase of carbon dioxide in the atmosphere will be offset by the fire, pestilence, droughts, floods, and storms that will come with it.

A warming of just 1.8°F (1°C) could threaten survival of plant species growing at the upper limit of their temperature range, particularly in the alpine regions in New Zealand and in southwestern Australia. Species that can't move or migrate because man has cleared the land, because the soil is different, or because there's a mountain or river in the way could become endangered or extinct.

Europe

The continent is a rich and stable region with well-developed economies, stable governments and political institutions, and good technical support systems. The urban environments will do better in a globally warmed world. The natural environments, which have been over-utilized for centuries, won't. Most natural ecosystems in Europe are managed, fragmented, or under stress from pollution or other human impacts.

Hot Debates

Europeans in general seem more concerned about global warming than much of the rest of the world, possibly due to the fact that if the global ocean conveyor belt, the Thermohaline Circulation, shuts down, Europe could be the big loser.

Europe has a total population of 720 million people. Though it does not have as many people as Asia, it has a lot less land. In fact, Europe has the highest population density of any other continent. It has the lowest birth rate and the most rapidly aging population of any other continent as well.

Potential impacts vary substantially between northern and southern Europe. Water shortages are projected for the south, while floods will hit the north. More marginal and less wealthy areas will be less able to adapt.

Climatoids

Europe has seen vast changes in precipitation in the twentieth century. Rain and snow have increased over northern Europe by as much as 10 to 40 percent, while at the same time southern Europe has dried as much as 20 percent. The number of rainy days over Spain's southern coast and Pyrenees Mountains has decreased by 50 and 30 percent respectively in just the last half of the twentieth century.

Forests

Most European forests are managed for timber, water, or recreation. This process has basically reduced or modified forests over much of the continent. Present day forests often consist of trees that are different from those that would have occurred naturally.

High summer temperatures and drought could increase insect populations and bring down those trees.

CAUTION   **Global Warnings** _____

European forests have stayed relatively isolated and semi-immune to pest organisms. The pine nematode, *Bursaphelenchus xylophilus*, which originates in North America, hasn't taken hold because of low summer temperatures and a short growing season, although the pine nematode frequently occurs in imported timber. If this and other pests were to establish a toehold in European forests, it could prove disastrous.

Fewer Attractions for Tourists

Higher incomes and more leisure time should expand the tourist industry in Europe, but global warming may change where the tourists go. The Mediterranean might lose a little of its popularity under increased summer heat, prolonged droughts, and forest fires. The coasts of Europe are a primary attraction for the tourist industry, but sea-level rise might damage buildings, bridges, and breakwaters. Sea-level rise could spell disaster for places like Venice, Italy, which is sinking as the sea is rising.

Lots of long, hot, polluted summers might make European cities less attractive. The mountains might get a bigger share of the summer business, but ski resorts might suffer in winter because of reduced snow. If the Alps increase in temperature by 5.4°F (3°C), the snowline will rise by 984 feet (300 meters), the first snowfall will be delayed, and below 3,927 feet (1,200 meters) there will be no continuous snow cover. Resorts might turn more often to artificial snow and Scotland's skiing industry might dry up altogether.

Latin America

The region of Latin America includes all of the continental countries of the Americas from southern Mexico to Chile and Argentina. Despite the fact that the annual population growth is declining from 1.68 percent in 2000 to an estimated .51 percent in 2050, that will still mean another 838 million mouths to feed. One of the major difficulties for this region is having enough food. Many countries already suffer from poverty, deforestation, and dwindling resources, and global warming will exacerbate these problems.

Storms

Flooding and landslides are common in Central America and southern Mexico. Strong weather and the inability of the land or the society to contend with it have produced numerous disasters. Latin America also has one of the greatest disparities of income in the world. The rich seem to get richer, and the poor get poorer here, and the poor often bear the brunt of major storms and changes in climate. The rich build their haciendas on dry spots up on hills. The poor throw up shacks in dangerous river valleys that are particularly susceptible to flooding. Hurricane Mitch created some of its worst havoc in river valleys in El Salvador and Nicaragua where raging waters wiped out whole towns.

El Niño and La Niña typically affect the various regions of Latin America dramatically. Some of the most severe droughts in Mexico in recent decades have occurred during El Niño summers. Then the Pacific side of Central America gets less rain than normal, while the Caribbean side gets more.

Whereas the enormous Andes Mountains get deluged in an El Niño, the Amazon Basin north to the Caribbean suffers from intense droughts and numerous forest fires. At the same time that the north of Brazil is burning, the south is under the same intense rains that are falling over the Andes. The principal danger with a globally warm Latin America is a permanent El Niño condition, with its extremes of weather and a reversal of conditions to essentially the opposite of what the economies, the societies, and the geography have become adjusted. That means rain in normally dry climates, and dry climates in normally wet ones.

Bugs

Latin America is famous for its wildlife diversity. The tropical forests that surround the equator, shroud the Amazon basin, and run north through Central America into southern Mexico have some of the most varied plant and animal species in the world. The same goes for their bugs. I remember going to Corcovado, a tropical forest in Costa Rica, one of the first Americans to visit after it had been declared a National Park. For my stay, I was told to bring "nets" and so packed a roll of cheesecloth, thinking that would do.

By the time I got to Corcovado, I had already traveled through Mexico, Guatemala, El Salvador, and Honduras, and my insect repellent was getting low. When I arrived there, I was shown to a large sleeping dormitory where there were no screens to keep out the bugs. Instead everyone there had full nets hanging from the ceiling surrounding their bunks like something you'd see in an English movie about Africa.

Different species of bugs arrived throughout the night, and you could watch the change of the insect guard around the light the camp kept on outside the building. It was too hot to wear blankets. I tried to surround myself with cheesecloth, but even that got hot, and I would kick it off in my sleep and the bugs would start eating the exposed flesh. When I woke up the first morning, I counted the bug bites on my left arm, which had dangled outside the cheesecloth. From my hand to my elbow were over 40 bites, and around my elbow there were so many I couldn't differentiate one from another.

Those bugs can sometimes carry disease. Most biologists that travel to Peru, Ecuador, Costa Rica, Guatemala, and southern Mexico get pumped full of shots, and then stop at the local "pharmacia" to fill up on antibiotics in case something gets past their vaccinations. The tropical jungle is beautiful but difficult.

With global warming, the bugs will be out in force. The mosquitoes will start swarming with the increase in warmth and humidity, as will the parasites and viruses within them. Scientists who've studied mosquitoes in Iquitos, Peru, find that a shift in climate of only 1.8–3.6°F (1–2°C) can have dramatic effects on the spread of mosquitoes and the diseases they carry. Mosquitoes that carry dengue and yellow fever have been reported at elevations of 7,220 feet (2,200 meters)—a lot higher than normal—in Columbia following periods of heavy rain. This increases the threat to people that live at high altitudes. Models project a global increase in the number of people at risk of malaria by 260 to 320 million due to an expansion of potential transmission zones. Increases in precipitation might drive cholera outbreaks as well.

The Polar Regions

Climate changes in the polar regions are expected to be the greatest of any area on Earth, and will cause major ecological, sociological, and economic impacts, especially in the Arctic and Antarctic Peninsula. The Arctic has warmed as much as 9°F (5°C) in the twentieth century already and sea ice has decreased by 2.9 percent per decade. The permafrost has begun to thaw in places, and this mucks up highways and slows traffic. In the Antarctic Peninsula a marked warming trend has caused spectacular losses to some ice shelves. Warming is likely to further reduce the ice, expose more bare ground, and eventually change the biology of the area. In areas of heavy permafrost, thawing will cause severe damage to roads and buildings.

There will be some benefits, such as a reduced demand for heating fuels. With a loss of sea ice, the Arctic Ocean may be open for more ships. And more ships will bring more commerce and more tourists.

The Winners and Losers

Indigenous people may be the least equipped to cope with changes in climate. Indigenous people number about 1.5 million of the total Arctic population of about 10 million people. They rely on the harvesting of fish, oil, minerals, and timber. These resources form the basis of their culture. The hunting of wildlife is an essential part of Inuit tradition. Hunting not only provides clothing and food, but also contributes to community cohesion and self-esteem. Knowledge of wildlife and the environment strengthens community relations.

The sea ice is vital to a number of forms of hunting. If the ice doesn't form, hunting is delayed or doesn't take place at all. Walrus, whale, seal, fish, and birds are dried in the spring and summer to preserve them for the long winter months, but when air is too damp and wet, food becomes moldy and sour. The length of the wet season also affects the ability of indigenous people to gather foods such as willow leaves, beach greens, dock, and wild celery.

Global Warnings

The permafrost in the polar regions is an important substrate to permanent structures. In fact, a 3.5°F (2°C) rise in the soil temperature in the Yakutsk region of northern Russia decreases the bearing capacity of ground under buildings by 50 percent. If the permafrost melts, a lot of architects and engineers in the polar regions will have to go back to their drawing boards to rethink existing structural designs.

Technologically developed communities in the polar regions are likely to adapt a lot better. As the oceans start to thaw, increased commerce and tourism will support changes in roads, buildings, planes, and cars. It will be easier for oil companies to build offshore oil wells. They won't have to worry so much about the ice. However, as the permafrost melts, it will muck up the roads and undermine the buildings.

There will be new sea routes opening in the Northeast and Northwest passages. The fisheries for salmon, cod, shrimp, and herring may increase. Ships will be able to use these areas without strengthened hulls. There will be new opportunities for shipping oil, gas, minerals, timber, freight, and tourists. But they'll need new harbors, new international maritime laws, and more collision insurance to protect from inevitable spills.

Driving Others

The polar regions contain important feedback mechanisms that can increase the warming. Many of these mechanisms are self-amplifying and, once triggered, will affect other regions of the world for centuries to come:

♦ Snow and ice reflect heat, but if warming shrinks this white cover, the newly exposed ground will absorb the heat and further reduce the ice.

♦ Global warming could shut off the global ocean conveyor belt, the Thermo-haline Circulation, with widespread and catastrophic effects.

♦ Projected warming will alter vast areas of oceans, wetlands, and permafrost, which act as major storage areas for CO_2 and methane.

♦ Reduced circulation in the Southern Ocean will decrease the ocean's ability to absorb excess carbon dioxide.

Small Island States

Some of the most concerned of the attendees at conferences studying or negotiating the issues of global warming are small island states. The Republic of Maldives—1,190 coral islets in the Indian Ocean southeast of Sri Lanka—has no tropical forests, hardly any automobiles, and little industry beyond the canning of bonito, yet its representatives are regular attendees at major global warming conferences. Their big concern is how much longer they will exist. Few of the islands have any point more than six feet above sea level.

Most of the island states of the world are located in the tropics or subtropics. They span the regions of the Pacific, Indian, and Atlantic oceans as well as the Caribbean and Mediterranean Seas. Although it is hard to generalize about these varied communities, they share some common features. They are all surrounded by the oceans, have limited natural resources, large population densities, and poorly developed in-frastructures. They also have limited funds, human resources, and skills. In general, they are sitting ducks for global warming and its attendant rise of cyclones and sea levels. And the effects would be dire indeed for coral reefs, water resources, agriculture, fish, and the tourism industry:

♦ Coral reefs, mangroves, and sea grass beds of coastal mainlands will be severely threatened by rising sea temperatures and sea levels. El Niño and its warming are notorious for bleaching out coral reefs. Changes in sea levels could drastically affect mangrove forests, which provide vital protection for tropical coasts, bays, and estuaries.

Global Warnings

The 200,000 people living on the 1,190 islands of the Republic of Maldives may be some of the first to suffer from global warming and its predicted rise in sea levels. Few spots on this Indian Ocean archipelago, with its 2,000-year-old culture, rise more than six feet above sea level.

◆ Water resources and agriculture are critical to small island states. Changes in the height of the water table and an increase of salt in fresh water from rising sea levels could be catastrophic for many crops.

◆ Tourism is the major source of revenue for many island nations. Almost all settlements, structures, and tourist activities are located along the coasts. Changes in temperature and rainfall, as well as the loss of beaches, could devastate the economies of many small island nations.

As you can see from the examples in this chapter, the effects of global warming would be devastating to many regions on our planet.

The Least You Need to Know

◆ Global warming will gift Africa with floods, droughts, famine, and disease.

◆ Asia will have more trees in the north and a few more crops in the mid-latitudes, but will lose land in the southeast delta regions.

◆ Australia and New Zealand will get hotter and drier, and lose plant and animal species.

◆ European cities and beachfront resorts will be hot and muggy in the summer, and the Alps won't get as much snow in the winter.

◆ Latin America will get buggier in the tropics and stormier in southern Mexico and Central America.

◆ The polar regions will sustain some of the greatest changes, and some of those effects could radiate out into the rest of the world.

◆ Small island states might sink under the rising seas.

Chapter 15

On the Home Front

In This Chapter

- ◆ North America gets toasty
- ◆ Floods but not enough water
- ◆ Trees on fire
- ◆ Lots more storms

Findings by the United Nation's Intergovernmental Panel on Climate Change as well as the U.S. Environmental Protection Agency suggest that under global warming scenarios predicted for the next 100 years, North America will be quite a different place than it is today. Our water, forests, wild lands, coasts, houses, dams, and our health will change as global warming begins to affect our climate. Let's take a look at some of the particulars.

North America Versus Global Warming

When the IPCC grouped countries in the western hemisphere, they referred to Canada and the United States as North America. They referred to Mexico, Central America, and South America as Latin America. Since this is the basis of their monumental study on global warming, I'll keep it that way for this book. No offense is meant to any of my Mexican friends—of whom there are many. My atlas says you are North Americans. You will always be North Americans to me. Don't blame me, the people at the UN Environment Program figured this out.

Like our Latin American friends, we're in for a hot time. Climate change will significantly affect natural ecosystems, where and how we farm, the availability of water to irrigate crops and produce power, the wetlands that spawn our fish, the beaches we use for recreation, and all levels of government and industry. Our water resources are heavily managed, yet they weren't designed for what the globally warmed weather might have in store for us. Insects and fire might play havoc with our forests. The rising seas will attack coastal areas. Storms will be brewing, and Florida might be sinking.

North America is a land of opportunity. Its gross domestic product is way up there. North Americans live long and spend lots of money. But they also use an enormous amount of natural resources and emit a lot of carbon dioxide. Though North America has a relatively clean environment, its use of fossils fuels for transportation and industry makes it the largest per capita producer of carbon dioxide in the world.

Global Warnings

Data from tree rings and lake levels in California show that in past millennia there have been severe droughts of more than 80 years in duration. Evidence also suggests that there have been long periods where the Rocky Mountains and Great Plains regions were much drier than today. These incidents happened without the added push of global warming. And they caution us against assuming that tomorrow's climates will be as benign as today's.

North America appears unready for the consequences. It's complacent. North Americans view climate as a constant. Oh, it might vary a bit from time to time—a few storms here, a few droughts there—but nothing to worry about. But scientists who've studied the climate find that it is anything but constant. Even without the added risk of increasing the greenhouse gases, the climate has varied greatly in the past. Over the course of the twentieth century there have been periods of extreme weather. The drought of the 1930s altered large portions of American lands and society. In recent years damage resulting from droughts, storms, flooding, hurricanes, and tornadoes has been on the rise.

Water

As the temperature warms, the ability of the atmosphere to hold more water will increase, and so will precipitation. But the picture won't be even. Climate change will affect different regions in various ways. The Southeast, Gulf, and Mid-Atlantic States might experience more floods, but in the West reduced snowfall and rising temperatures will increase the risk of drought. With less snow, there will be less snowmelt to feed important water sources like the Colorado and Columbia Rivers. With higher temperatures and less snow, most of the flows could take place in winter and spring, leaving the summers, when snowmelt normally provides water reserves, much drier than normal.

Floods

The ability of the warmer air to hold more moisture should lead to an increase in floods in some areas. Subtropical conditions will extend further north into the United States, which will change the precipitation, the vegetation, and the potential for disease in the southern states.

But with warmer temperatures come warmer rains and less snowfall. In mountainous areas of western North America, small high lakes and snow banks contribute the bulk of the flow of many river systems. Warmer winter precipitation will not only increase winter and spring flows but will also increase the chance of rain on snowfall—a combination that provides some of the greatest risk of floods. In the Southwest, increased rains may also lead to increased flash floods, a result of the fact that the desert landscape has so little vegetation to hold back moisture that it can't handle rapid increases in runoff.

The water resources of North America have been heavily modified to increase the long-term ability of man to use the water. Private, state, federal, and provincial governments have built everything from private ponds to major dams. Water management practices like these have made it feasible for most North Americans to have access to water both in years of high and low flows. There has been a price to pay. Dams and levies often radically alter the natural ecosystems. In some areas, dams are being removed or considered for removal to restore native fisheries.

The human-made reservoir capacity in North America is equal to approximately 22 percent of the average annual runoff. That compares to an average of 10 percent worldwide. In the United States alone there are 75,000 dams, with a combined storage capacity equal to 70 percent of the average annual runoff. But storage capacity was built on past water records. Toss in a flood or a drought and you might get a system that still won't meet increased demands. Plus levees and dams are often successful in managing most variations in the weather, but when precipitation gets heavy and the dam breaks, the damage can be a lot worse than if there had been no dam at all. One community's efforts to protect itself from floods through the construction of dams or levees may just end up increasing the likelihood of flood damage to another community downstream.

Global Warnings

Heavy rains can cause radical disturbances to the land and destroy freshwater and marine environments. The rainfall from Hurricanes Dennis and Floyd in eastern North Carolina in September 1999 provide an example. Rainfall of 3 feet (0.9 meters) breached containment ponds for poultry and hog waste and turned the flooding waters into a soup of raw sewage, fertilizers, and decaying vegetation. The floods contaminated the ground waters and eventually reached the waters of the estuary between the Carolina mainland and the Outer Banks and turned the water brown. That area is one of the main nursery grounds for commercially important fish from Maine to Florida.

Droughts

Though some areas may get increased precipitation, global warming may subject the Midwest of the United States and the Canadian Prairies, the agricultural heartland of the continent, to severe droughts. The natural environment might actually shift from grassland to something a lot closer to desert.

Heat not only reduces snowfall but increases evaporation. The latter may reduce the water levels of the Great Lakes and the outflow into the St. Lawrence River. Reduced snowpack and increased evaporation might also lead to reduced stream flows and lake levels in many parts of Canada, despite the fact that scientists predict increased annual precipitation. Declining snowpacks and increasing temperatures will essentially counteract the effect of more rain.

This is the problem in the West, too. Declining snowpacks will lead to heavier winter and spring rains and floods, but decreased summer snowmelt, which will lead to decreased summer water. This will affect both wildlife and agriculture. In the arid Southwest, this could lead to increased reliance on groundwater and a lowering of the water table. The end result could be that more water is moved from agriculture and put into urban use.

Your Water Bill, Sir

Will all these water problems increase your bill? Is the Pope a Catholic? Even in the face of hopeful increases in the conservation of water, the projected increases in water demands for the growing populations would cost $13.8 billion more in 2030 than in 1995. This is in terms of dams, aqueducts, and the purchase of water rights. And we're not talking global warming here. This all comes from the high costs of developing new water supplies, environmental concerns, a growing appreciation of natural stream-flows, and efforts to improve water quality. Now add some global warming to the equation, and the projected annual increase will go from $13.8 billion to $105 billion—about $308 per person annually. A family of five will pay about $1,540 a year more for water than they do today.

Forests and Fires

North America contains about 17 percent of the world's forests. Temperature and precipitation strongly affect the type and density of forests. Climate change at pre-dicted levels could eventually move the southern boundaries of most North American forests as much as 400 miles (600 to 700 kilometers) northward. This is because the southern limits of forests will become too hot for individual forest types and there will be massive die-offs. The northern boundaries, however, would move only as fast as the normal rate of migration for forests. Projecting a migration rate of 60 miles (100 kilometers) per century—about double the known historic rate—the ranges of forests might shrink, simply because the southern boundaries may advance more rapidly than the northern perimeters. Forests might climb higher in mountain environments though at the expense of increasingly rarer arctic-alpine meadows and tundra. In general, the north *boreal forest* should advance poleward as the permafrost unfreezes.

Warm Words

The **boreal forest** is the most northern type of forest and is dominated by coniferous trees such as the evergreen spruce, fir, and pine, and the deciduous larch or tamarack.

Global warming should be easier on managed forests, because humans can help the forest adapt to change. But it will be harder on native forests that don't have that added man-made assistance. If climate change results in the wholesale replacement of one forest type for another, wildlife habitat and natural ecosystems could be altered dramatically.

Global Warnings

> Global warming might significantly alter forest types and reduce the land area of healthy forests. A study of forests in northern Mississippi and northern Georgia indicates that seedlings currently in such areas would not grow due to higher temperatures and drier soils that global warming would bring. In central Michigan, grasslands with some sparse oak trees might replace forests now dominated by sugar maple and oak. In northern Minnesota, the mixed boreal and northern hardwood forests could become entirely northern hardwoods. Changes could begin in only a few decades.

Drier soils expected to accompany global warming could lead to more forest pests and diseases. Add heat, dry soils, and some extra pollution to a forest and you have the perfect combination for a forest fire. A lot of this depends, however, on the amount of warming. If there is only a small amount, then the increased warmth and increased atmospheric carbon dioxide could actually promote forest growth. But as the warming continues to rise, increased water use associated with higher temperatures overwhelms the carbon dioxide effect, resulting in fewer trees.

Tourists Go Home

The United States and Canada are among the top 10 tourism destinations in the world with 1998 receipts of $71 billion and $9 billion respectively. However, tourists are apt to visit a place quite different than the ones the postcards and the coffee table books depict. Parks and other natural areas are important tourist destinations based to a considerable extent on the species they preserve and the natural environments they sustain.

One study revealed that 75 to 85 percent of Canada's National Park system would experience major shifts in the dominant vegetation under global warming scenarios. Large areas of the Yellowstone National Park region would experience extinction of a number of plants.

Hot Debates

Golfers may be some of the biggest losers in a globally warm world. Increased municipal and agricultural demands for water might inch out plans for developing new courses, and in times of drought, diminish the capacity to irrigate existing facilities economically. Even though the warmer weather may increase the recreational season, the deleterious effects that global warming might have on recreational lands will spoil the extended party.

Then there are your beaches. Coastal zones are among the best tourist destinations in North America, accounting for up to 85 percent of all tourist dollars in the United States. But sea-level rise will wreak havoc, particularly to beaches backed by seawalls or development that can't migrate backwards. One study estimates that just to replace the sand on U.S. beaches that would be lost to a 20-inch (50-centimeter) sea-level rise would cost $14 to $21 billion.

Farmers beware.

(Photo by Michael Tennesen)

Agriculture

In most areas of the United States, global warming will reduce yields of corn, wheat, and soybeans with individual farm losses ranging from small amounts to 80 percent of the crops planted. The decreases would be primarily the result of higher temperatures, which would shorten a crop's life cycle. In Canada and the northern portions of the United States, yields of corn and soybeans could increase as warmer temperatures extend the frost-free growing season. Southern areas growing heat-tolerant crops such as citrus fruit and cotton might also benefit from a reduction of killer frosts. Still yields may decline in southern Florida and Texas because of excessive heat during the winter.

In general, global warming and increases of carbon dioxide in the atmosphere will work both

Global Warnings

And then there are the bugs. Higher temperatures and warmer winters could reduce the winterkill of insects as well as broaden insect ranges. Toss in a lot of rain and that tends to reduce the efficacy of pesticide applications. Plus, it can move pesticides into areas where you don't want them.

ways, making some crops grow while others wither on the vine. Increased yields in the mid- and high latitudes may come as a result of the positive effects of carbon dioxide (remember plants breathe that stuff), a longer growing season, and a lessening of cold temperatures. On the other hand, decreases in yields in drier areas will come courtesy of a lack of water.

An increase in storms won't help either. One of the major reasons that the Corn Belt region of the United States is so productive is that temperatures don't vary that much from year to year across the region. The main risk of climate change to some regions of North America may be primarily from an increased variability. Corn, wheat, and soybeans know what they like, have grown accustomed to the weather on the local farm, and, like many of us, don't like a lot of change. Climate change, however, appears to be on the horizon.

Ranching

Global warming may increase livestock appetites while it alters grasslands and feed. Most estimates suggest that the negative effects of hotter weather in summer outweigh the positive effects of warmer winters. An expected increase in storms could breed havoc, though. The severe ice storms in eastern Canada and the Northeast United States in the winter of 1998 were brutal on livestock in those regions.

Fisheries

Variations in the climate play a big role in a number of North American coastal and deep ocean fisheries. Changes in sea temperature, ocean water nutrients, and circulation patterns that are expected with most global warming scenarios may affect the populations of various species of fish. Climate variations may reduce the abundance of microscopic plant and animal life in the sea, and this can cascade through the chain of marine predator-prey relationships.

Fisheries management is going to be a rough job in the presence of climate change as managers try to figure out where the fish are, how many they have, and who is entitled to what. Recent history is not on their side. The Pacific Salmon Treaty was embroiled when declining runs of southern coho and chinook salmon were coupled with the increased abundance of salmon in Alaska. That threw a monkey wrench into efforts to achieve a mutually acceptable balance of U.S. and Canadian interceptions of one another's salmon. The collapse of the cod stocks off Newfoundland on Canada's east coast is another example of the inability of government to control a fishery when coupled with less favorable environmental conditions.

Human Health

Human illnesses and deaths are variously linked to weather patterns. Bad weather brings greater incidence of flu and pneumonia. It can also whip up the pollen, which has effects on asthma and allergies. Death rates increase for the sick and the elderly when extreme weather arrives. Increased temperatures and rain also affect disease-carrying mosquitoes and ticks. They bring Lyme disease, Rocky Mountain Spotted Fever, and dengue and yellow fevers. Global warming might also bring an increase in pollution, which can heighten the incidence of asthma, emphysema, and other respiratory diseases.

Frosts and Heat Waves

In a warmer world, heat waves are expected to be more frequent and severe. And that can cause increases in illness and death, particularly among the young, the elderly, the poor, the ill, and those who live on the top floors of apartment buildings without access to air-conditioning. Still more people die of exposure to cold than exposure to heat. So, since there will be fewer cold waves, there should be a decrease in weather-related deaths.

> **Global Warnings**
>
> In the United States, populations in Northeastern and Midwestern cities may experience the greatest number of heat-related illnesses and deaths based on past experience. Heat waves in urban areas in these regions have been brutal. Heat killed 118 people in Philadelphia in 1993, 91 persons in Milwaukee in 1995, and 726 persons in Chicago in 1995.

Malaria, Anyone?

Malaria is today uncommon in the United States and Canada. But it wasn't always that way. As recently as 1890, there were more than 7,000 malaria deaths per 100,000 people in the United States and southern Canada. In Canada, the disease disappeared at the end of the nineteenth century. By 1930, malaria was under control in the Northern and Western United States and generally caused fewer than 25 deaths per 100,000 in the South. In fact, in 1970, the World Health Organization's Expert Advisory Panel on Malaria recommended that the United States be included in their register of places where malaria had been eradicated.

Malaria is still found in tropical and subtropical regions. But a number of studies indicate that with global warming, malaria may drift northward from its current tropical residence and be reintroduced into the United States and southern Canada. Public health measures will play a large part in the existence or spread of the disease in a globally warm future.

Storms

Have we got some storms for you? The U.S. Federal Emergency Management Agency (FEMA) declared fewer than 20 natural disasters annually in the 1950s and 1960s but more than 40 a year in the 1990s. Global warming could up that ante. We've already talked about floods. There will be lots of them, especially in winter and spring. Sea-level rise could increase the tidal surge damage that will come from hurricanes. Scientists are up in the air as to whether we'll have more or fewer of these monsters. But they do agree that the ones that come ashore will be pushing more water ahead of them and bringing more rain along as well.

Snowstorms might be down, but areas susceptible to freezing rains will increase. These severe winter storms are rough on power transmission, and the combination of severe winter storms and loss of power can have devastating consequences even in highly developed regions such as Ontario-Quebec and the Northeastern United States.

In one such storm in January 1998, more than 31 inches (80 millimeters) of freezing rain—double the amount of any prior ice storm—produced the largest estimated insurance loss in the history of Canada. Power poles and power lines gave in under the weight of all that ice, stranding some residents and farmers without power for as much as four weeks. Almost five million people were without power at some point during that storm. Total Canadian losses were about $4 billion in U.S. currency. With global warming, ice storms will become more common.

Climatoids

An example of the type of storm manipulation that the world might become more used to in a globally warm environment is the cloud-seeding program underway in Alberta, Canada, since 1996 to reduce the severity of hailstorms. Central North America is one of the most hail-prone regions of the world, with hailstorms accounting for more than half of its catastrophic weather events. A 1995 hailstorm did more than $1.1 billion damage in Dallas, a 1990 storm did $600 million damage in Denver, and a 1991 hailstorm did about $240 million ($340 million [Canadian]) damage in Calgary.

Wetlands

About 14 percent of Canada's surface area is covered with wetlands, about 24 percent of the global total. About 6 percent of the United States is wetlands. Human activity has greatly affected mid-latitude wetlands over the last 200 years. Agriculture and development have destroyed more than half of these vital areas in the United States. Global warming will affect what's left.

Sea-level rise will alter coastal wetlands in many areas with potential damage to ocean fisheries. Increased drought conditions in the Prairie Pothole Region of the northern Great Plains will have drastic effects on U.S. duck breeding populations.

You Wouldn't Want to Live by the Coast

Some 65 percent of people in North America live in coastal communities. This includes people who live near the Atlantic and Pacific oceans, but also those who live near the Great Lakes. Everybody is going to see some changes. The worst will be for those on the coast who may receive an occasional combination punch of storm surge and sea-level rise. As the seas rise, salt water will work its way into estuaries and other aquatic systems with injurious effects to wildlife, agriculture, and industry. High seas and more floods will make it increasingly difficult for those who live in coastal bays and river deltas.

Coastal ecosystems are clearly vulnerable to changes associated with sea-level rises and increased runoffs. The increased frequency in hazardous blooms of algae in estuaries in the Gulf, Atlantic, and Pacific states has threatened both the commercial fishing industry and the public health. Consumption of toxic algae by shellfish can create concentrations of toxins in the shellfish that are high enough to cause paralysis and death to humans unlucky enough to eat them.

Got Insurance?

As shellfish poisonings are piled on top of ice storms, catastrophic floods, and storm surges, the tab is going to run up. Recent increases in financial losses from hurricanes, winter storms, and floods have raised questions about what it's going to be like when we toss in some global warming. In the United States there were 29 events from 1980 to 1997 where losses exceeded $1 billion.

The most costly were the droughts and heat waves of 1988 and 1990, Hurricane Andrew in 1992, and the 1993 Mississippi River flood, which had combined costs of $100 billion. In Canada, the 1996 Saguenay flood and the 1998 ice storm in Ontario and Quebec cost more than $1 billion. Make sure you are covered.

Florida Goes Down with the Ship

The prospect of sea-level rises from 4 inches to 3 feet (.09 to .88 meters) over the next century is one of the most dangerous impacts of climate change. The greatest impact will likely be in recently developed coastal areas like Florida and parts of the U.S. Gulf and Atlantic coasts. Insured property value in Florida alone exceeds $1 trillion.

In Florida, sea-level rises of only a few inches will, because of the gradual slope of the state's coastal areas, advance the shoreline greatly. In some cases the horizontal advance will be 150 times the vertical rise.

Sea-level rise may advance as much as 400 feet in low-lying areas. This will flood shoreline homes and hotels and erode the state's treasured beaches. Sea-level rise might also damage the Everglades, offshore islands, and coral reefs. Global warming might also increase forest pests, which will make the trees ripe for fire. In short, much of what has made Florida so precious might be lost in a globally warmer world.

The Least You Need to Know

- Global warming will bring North America floods, droughts, and high water bills.

- The forest might benefit or might shrink depending upon the amount of warming.

- Parks, beaches, and other tourist attractions will suffer as the things that draw the tourists are destroyed.

- With global warming we might get deathly sick, but at least we won't freeze to death as often.

- Storms might become a regular part of the human budget.

- Sea-level rises might take their toll on developed coastlines like Florida and other Gulf and Atlantic states.

Chapter 16

Nature Coughs Up Her Share

In This Chapter

- ◆ Polar bears get hungry
- ◆ Bighorn sheep get chased
- ◆ Songbirds get scarce
- ◆ The place is crawling with fire ants

If the climate did get warmer, animals could move north. They've done it in the past. When the great glaciers of the ice age advanced south and later retreated north, the wildlife followed them back and forth. In fact, Darwin in the 1850's asserted that if "they all migrated in a body together, their mutual relations will not have been much disturbed." But the transition from the ice age to the present age, the *Holocene*, took thousands of years, during which time the average global temperature increased 9–12.5°F (5–7°C) and the sea level rose 328 feet (100 meters). However, what we may do in the next 100 years is to increase the speed of that temperature change by about 50- to 100-fold. It's not that same gradual increase. Things won't be so smooth. Some transitions are occurring right now and they don't look choreographed.

Warm Words _____

The **Holocene** is the epoch of earth history dating from about 10,000 to 11,000 years ago, up to the present.

Mammals

Climate changes have had enormous effects on wildlife in the past. Younger Dryas, the big dip back into the ice age that happened at the beginning of the Holocene, might have been a factor in the extinction of the great ice age mammals—the mammoth, mastodon, native camel, ground sloth, giant bison, and the saber-toothed tiger. The coming age of global warming could have as drastic an effect, causing wholesale extinctions across the globe. Some effects are already here.

Polar Bears and Ringed Seals

Every year around December, about 200 to 250 female polar bears give birth to twins or triplets in the Wapusk National Park area. They are part of a population of about 1,200 polar bears in and around the park, some 600 miles north of Winnipeg in northern Manitoba. The females fast during their first months with their cubs, but as spring approaches, they grab their cubs and make a run for it out onto the ice. The females desperately need food and must reach the Hudson Bay before the ice breaks up and ringed seals, the favorite food of the Wapusk National Park polar bears, disappear. The seals come here to mate and bear their young. It will take a family of polar bears about two to three weeks to make it to the ice, during which time each female will probably lose one or more of her cubs. It's a tough journey.

Once the family reaches the ice, they'll feast on seals. Mom will teach whatever number of cubs survive how to hunt themselves. If the cubs get through their first months outdoors, they'll spend the next two years by their mother's side, migrating to and from the Arctic ice packs and learning to hunt the seals. Polar bears gorge themselves during the seven to eight months of the year when the ringed seal is available. During that time they can triple their body

Climatoids _____

In summer when the breakup of the ice on the Hudson Bay forces polar bears to abandon their favorite prey, the ringed seals, the polar bears move inland onto the summer tundra. There they enter a walking hibernation as their metabolism slows dramatically to conserve energy. They then feed only on occasional lemmings, voles, and some vegetation.

weight. Males may grow to 1,000 pounds or more during their feeding period while females may reach 600 pounds. They'll need the extra weight to get them through the summer when the ice breaks up on the Hudson Bay, and the polar bears move inland onto the summer tundra.

Polar bears are the world's largest land-based predator. They have only one enemy, traditional Inuit natives who hunt the bear for their meat and fur. But now they have another enemy, human-caused global warming. Today, ice melts in the Hudson three weeks earlier in the spring than it did just 25 years ago.

The increasing sun and early ice melt means there is less time for the polar bears to fatten up for the summer months when they won't have access to the seals. Polar bears amass most of their body fat during their spring feast of ringed seals. When the ice disappears earlier, they aren't able to put on as much fat. That fat is critical for the males, but even more critical for the females who must fast an additional period while they nurse their cubs. Females lose an enormous amount of weight giving birth and caring for their cubs, but gain it back if they have enough time on the ice. One female weighed only 250 pounds after her pup's birth, but ballooned to 880 the next year.

But scientists have noted a 10 percent drop in the number of cubs born in the last 20 years. Adult bears are also 10 percent thinner. At a lower body weight, females have a tougher time nursing, which is part of the reason there are fewer cubs.

Female polar bears use snow dens to give birth to their cubs. But rising temperatures can affect these dens themselves. If the dens melt early, the cubs are exposed to the outside world too soon. The same is also true for the ringed seals, which are the prey of the polar bears. They use snow dens to shelter their young. Without the snow dens the health of both predator and prey are affected. The Arctic ecosystem is one of the fastest-warming places on earth. The ice pack, essential to the survival of many species, is rapidly shrinking.

> **Global Warnings**
>
> As global warming has increased the length and intensity of the summer sun near the Hudson Bay, it's afflicting the skin of Inuit children with a condition many mothers have never seen before. It's called "sunburn."

Caribou

Global warming could affect caribou, also known as reindeer. Presently the caribou are distributed over much of the higher latitudes of the northern hemisphere. Reindeer run more than any other deer. They protect themselves by gathering into large herds and shaking off predators through fast and tireless runs. When wild caribou get frightened, they follow the first animal that breaks away from the herd. Though it is usually the older, stronger animals that lead the herd, a frightened cub bolting for its mother will sometimes take the whole herd with it. The Old World reindeer and the North American caribou belong to a single species, which has several forms and sizes. Males weigh anywhere from 216 to 605 pounds (125 to 275 kg) depending upon whether they live on the tundra (the skinny ones) or in the forest (the fatter ones).

During the summer, caribou wander continuously from one meadow to another in search of plants, fungus, and lichen upon which they feed. In some areas their wanderings might take them a thousand miles in one direction. As soon as the snow melts, they wander into colder areas, especially where the winds blow. These long migrations are not only in search of food but for windy places where they can get away from biting insects—bot flies, horseflies, gnats, black flies, and mosquitoes. They suck a surprising amount of blood from the reindeer. The animals never really recover fully from these attacks until August when the flies begin to die back. Global warming would increase the number of insects. It could also lead to the spread of anthrax and hoof-and-mouth disease. There is a chance, however, it could increase food supplies in some areas, if the caribou can stand the flies.

Hot Debates

One point of contention between oil drillers and environmentalists has been the amount of disruption that is caused by oil pipelines to natural movements of caribou in the Arctic Circle. According to biologists, when caribou run across an unexpected obstacle on their route, such as a new gas pipeline or a frozen river, they can take as much as a month to figure out how to get around it. Meanwhile, they are subject to starvation and predation from the packs of wolves that follow them. Global warming is apt to increase this disruption by altering the terrain.

California Sea Lions Take a Dive

The California sea lion is another loser in a globally warmer world. This is one of the best known seals because it is easily trained for the circus and is frequently kept in zoos. True California sea lions are found off the shores of southern California, the coast of Baja California, in the Sea of Cortez, and off the Mexican Coast. California sea lions are noisy animals. Their voices have a high pitched bark, which is very distinctive.

I accompanied biologists one summer to San Miguel Island off the coast of California, a summer rookery for the animal. There they come to mate and bear their young. Females gather into harems and big bull males challenge each other in open-mouthed fights for access to the females. Males typically wear large gashes in their coats during the summer mating season. On San Miguel there can be up to 20,000 sea lions and other seals during the California sea lion mating season, and the noise is amazing. It's like walking into a big stadium and somebody has just made a touch-down, only the cheering goes on 24 hours a day.

The diet of California sea lions consists chiefly of squid and some other species. The calves are born during the month of June. The mother returns to the water to fish soon after the pups are born to feed their hungry mouths. Sea lions prefer to hunt in shallower waters both by day and night. The waters off San Miguel are rich with food, and in good years, about 75 percent of the pups make it.

But in the El Niño year of 1997, it was a different story. When the waters grew warmer, the normal prey species began to swim deeper and move further north. The result was catastrophic. In non-El Niño years pup mortality runs about 25 percent, but that summer it went up to 70 percent. With global warming presenting the possi-bility of a permanent El Niño, this catastrophe could become a regular occurrence.

Grizzly Bear and White Pine

The two last remaining grizzly bear populations in the lower 48 states reside in northwestern Montana and in and around Yellowstone National Park in Wyoming. In all there are less than 1,000 grizzly bears left where 100,000 once roamed over the western portion of the United States and Mexico. The grizzly bears are in a perilous situation. They are currently protected under the Endangered Species Act. Habitat and foods are important to their continued existence in these environments.

The problem is, their favorite food is white-bark pine seeds. The seeds are much larger and more durable than the seeds of other cone-bearing trees, and it requires much less energy to get at them. Whitebark seeds are usually ripe around mid-August, at which time a host of birds, chipmunks, and other animals including the grizzly bear harvest the cones en masse. Bears at this time go after the whitebark pine seeds almost exclusively, espe-cially adult females who must fatten up quickly if they are to produce cubs during the winter.

Climatoids

One of the grizzly bear's favorite methods of getting enough seeds is to let red squirrels do the harvesting for them. Red squirrels gather whitebark pine cones directly from the trees and store them in caches that grizzly bears love to raid.

The problem is that the whitebark pine is a sub-alpine species that exists in Wyoming at elevations from about 5,000 to 10,000 feet. As a warmer climate drives other species of trees higher, the whitebark pine could get squeezed out. Plus, the whitebark pine is subject to a number of diseases that could be affected by rising temperatures. Chief among them is blister rust. This fungus has already taken a toll throughout the range, as much as 90 percent in some areas. Mountain pine beetles also take their toll on whitebark pine trees.

Recent surveys show that blister rust infections in Yellowstone are still relatively low. Scientists estimate that only about 5 percent of the trees are infected. Biologists thought the area was unfavorable to blister rust, but global warming is making Yellowstone's environment more blister rust–friendly. Blister rust could dry up trees and make them more susceptible to fires.

The Sheep and the Lion

Desert bighorn sheep range over the Southwest from Texas to California and into Northern Mexico. There are about 20,000 of these animals in the United States today, a fraction of their former population. Bighorn sheep emerged from earlier evolutionary forms during the last ice age, when the glaciers stood at their southernmost boundaries. The sheep were then confined to an area of the once-lush southwestern United States. As these ice walls retreated, vast expanses of grassland grew up in their wake. The bighorn passively followed the endless grasses and thrived.

Climatoids

Bighorn sheep are famous for their highly stylized fights between males. Two rams who haven't decided who's king will often feed within fighting distance seemingly unaware of each other. Suddenly, both will rear up on their hind legs and begin churning rapidly toward one another. Bodies propel forward and down, adding gravity to the charge; heads flick forward at the last second and horns meet with a bellowing crack that can be heard more than a mile away. Often the two gladiators achieve a combined speed of 45 mph (72 kph) in the charge. Evolution has given the bighorn a double-layered skull, cross-connected with bone and thick facial hide, which enables the animal to withstand these tremendous forces.

As the Southwest became more arid, the desert bighorn adapted to an environment that sometimes reached 130 degrees. The desert bighorn can survive without water for months on end, getting most of their water by eating vegetation. They are normally secretive animals. It's hard to find one. There are tricks, though. Biologists showed me a watering hole near Lake Mead on the Colorado River above Boulder Dam and told me to come back sometime when temperatures exceeded 110°F (43°C) three days in a row and I would see bighorn.

The bighorn have excellent vision, equivalent to a pair of binoculars, so I was told not to try and hide. Instead, just sit out in the open about 50 yards back of the water hole, with a beach chair, an umbrella, and a cooler full of cold drinks, and see what happened. Sure enough, a group of more than 10 bighorn sheep came to the water hole a little after noon in the heat of the day. They hesitated for a while, but deciding the clown under the umbrella was obviously harmless, came in and drank and let me take pictures.

Bighorn sheep at peril.

(Photo by Michael Tennesen)

The bighorn like the high, rocky precipices where predators cannot follow. They have a unique double-shelled hoof that spreads over sharp, broken rocks, and a soft, cushiony pad in the middle for traction on slick surfaces. The sheep are capable of nimbly bounding up and down incredibly sheer mountain cliffs, often dropping 20 feet between contact points. They can also follow two-inch-wide trails across steep cliffs.

But desert bighorn are reluctant to leave their high, rocky refuges and move to flat lands between ranges where they can't outrun predators. Biologists literally have to pick them up and take them down the road to get them to spread into new territories. This makes them uniquely susceptible to climate change, since the animals that will survive best are the ones who can adapt and move.

However, bighorn susceptibility is more complicated than that. The bighorn are running into increasing problems with mountain lions. There are several populations in the Southwest where mountain lions have almost wiped out the desert bighorn.

In the early 1990s, when desert bighorn populations were plummeting in southern California mountains, California Fish and Game biologists radio-collared 100 animals to see if they could pinpoint the problem. When the results were tabulated, they discovered that over 70 percent of the bighorns that died in the study were killed by mountain lions. This is not the only instance where mountain lions have proved the nemesis of bighorn sheep. Mountain lion kills have reduced bighorns in the Sierra Nevada Mountains to less than 100 animals. In the San Andreas Mountains in New Mexico, hungry lions have eaten the herd down to a single female.

Eric Rominger, a professor at New Mexico State University, who works with New Mexico bighorn, admits the idea of the lions driving a herd into biological extinction goes against the classical predator-prey relationship in which the two populations are supposed to stay in equilibrium. In other words, as the prey dies back, so do the predators. But he claims that fire suppression policies have allowed forests to invade the normally bald rocky areas where bighorn are found. The trees then provide habitat for deer which are not usually found in dessert bighorn habitat. This allows resident mountain lions to prey on deer and stay healthy even though desert bighorn are declining. In an environment without the deer, the mountain lions would normally decrease as their prey decreased, so that they would never be in a position of wiping out a prey animal. But in this special situation, man has disrupted the environment and made "wipe-out" a clear and present danger.

In a globally warm world, trees will ascend the mountains and create havoc for desert bighorn sheep as well as for a number of other animals. This shows how complex some of these natural biological relationships are and how man-induced warming could throw them out of whack, with results that aren't expected.

Sea Birds

Adélie penguins are already feeling the effects of global warming in the Antarctic. Scientists studying the birds there find that only 1 out of 10 Adélie penguins survive the winter and return to nesting grounds to breed. They believe that a 5.5°F rise in sea temperature since the 1940s is the villain. The temperatures have caused the numbers of krill, an important food for the penguins, to plummet.

The 1997–1998 El Niño has not only affected California sea lions, but it also proved devastating for California's endangered brown pelican. In the spring of 1998, researchers found only 280 nests in a colony in the Channel Islands National Monument off California, which normally contained 20,000. Climate change produced by El Niño had sent the brown pelican's normal fish prey deeper and further north.

Pelicans may lose.

(Photo by Michael Tennesen)

Waterfowl

More frequent and severe droughts could dry up many of the prairie potholes, small wetlands which provide breeding habitats for most of North America's waterfowl in the Great Plains and southern Canada. Models developed by scientists at Boston University suggest that within 50 years climate change could wipe out 54 percent of the prairie potholes and about 60 percent of the duck population.

Songbirds Stop Singing

Climate change could cause major shifts in the range of many songbirds and some local extinctions. The forests of northern Minnesota and southern Ontario could lose up to 14 species of warblers. Warblers live off insects like spruce budworms, which ravage local forests, and their absence could lead to devastating insect infestations.

The Kirtland's warbler in northern Michigan provides an example of how global warming effects on one natural system will challenge another. The bird is confined to a narrow area of jack pines that grow in sandy soil. Models of global warming's effects on the system predict that the jack pine is likely to migrate north with the warming. The trouble is Kirtland's warblers are not likely to survive the transition. That's because the birds like to nest on the ground under young pines. But the soil to the north is not sandy enough to allow proper drainage for the nests. Without that drainage, water would accumulate and kill the nestlings. Scientists who study Kirtland's warbler predict that global warming could wipe the birds out within 30 to 60 years.

> **CAUTION**
>
> **Global Warnings** _____
>
> The combined effects of the fragmentation of wildlife habitat and warming present one of the potentially most dangerous results from global warming. During the ice ages many plants and animals survived by moving with the changing climate to another better-adapted area. But migration is not an easy deal in the twenty-first century, since wildlife must migrate across freeways, farms, industrial parks, military bases, and cities.

No Breaks for Reptiles and Amphibians

Reptiles and amphibians have a rougher time of it than birds, mainly because they can't migrate with the weather, as birds can. They are extremely susceptible to climate warming, which might be one of the reasons that amphibian numbers have already fallen around the world. Reptiles and amphibians are ectotherms, meaning their body temperatures are not regulated by their metabolisms like you and me, but by the sun and the outdoor temperature. Scientists studying the desert tortoise find that tortoises come out of their burrows earlier or later depending quite literally on the temperature. And they often have to warm themselves up on a hot rock before their externally regulated metabolism is warmed up enough to eat. When the temperature gets too cold in the winter, they go into hibernation and stop eating.

> **CAUTION**
>
> **Global Warnings** _____
>
> One study found that if we fail to do anything about global warming, habitat loss for salmon and trout could be as high as 17 percent by 2030, 34 percent by 2060, and 42 percent by 2090.

Will fly-fishing be another casualty of global warming? Good trout or salmon streams must be clear, free-flowing, and, very important, they must be _cold_. But global warming could change the chilly streams that sustain these prized fish. Roughly 10 million anglers spend about 10 days a year fishing for salmon and trout. The estimated value of these two fish is about $1.5 billion to $14 billion annually. Salmon are also an important part of the cultural heritage of northwestern Native Americans.

Why are trout and salmon so susceptible to global warming? That's because these fish thrive in streams that are 50–65°F (10–18°C). In many areas of the country, these fish are already living at the upper limits of their temperature range. That means that even small changes in temperature could render streams uninhabitable.

Insects

Butterflies such as the sachem skipper butterfly in the Pacific Northwest and the Bay checkerspot butterfly in California are already responding to climate and weather changes. This could foretell changes in other species that occupy the same ecosystem as the insects.

We've already discussed how mosquitoes and ticks, which carry potentially deadly diseases, will increase their ranges into North America as the climate gets warmer and wetter. Red imported fire ants may also do the same.

If you live anywhere around Florida, North Carolina, Oklahoma, and Texas, then you probably know about fire ants. Fire ants are stinging insects that came here from Brazil, and which in many areas of the Southeast have taken over ground from native ants. Fire ants make it impossible for many people in the Southeast to walk barefoot in their own backyards. Fire ants will attack farmers who've stopped to fix a plow, or even livestock trying to get to feed. Global warming may help these ants spread to other parts of the country.

Coral Reefs

Global warming could wreak havoc on the world's coral reefs, especially those already under stress by pollution, physical destruction, and over-fishing. If sea levels rise, that could submerge coral reefs too deep for their survival. If carbon dioxide concentrations increase in ocean water, that could inhibit the ability of coral to build up their protective skeletons. Warmer water also bleaches the coral by killing off the algae that gives the coral reefs their color. This bleaching makes coral weaker and more susceptible to disease, predators, and storms. Coral diseases are already on the rise around the world from warmer waters and increased levels of dust from droughts.

Tide Pools

Global warming could alter the habitat for Pacific Northwest tide pools, which are an important marine resource. If global warming creates more warm water in coastal intertidal zones, this could cause ochre sea stars, an important predator, to prey more heavily on mussels. The increase in predation on the mussels could upset the delicate balance of nature in this delicate system and produce damaging side effects that could ripple out into the deeper ocean.

Our National Parks

The animals and plants of our national parks are in trouble if the climate warms and shifts vegetative zones northward. Plants can't pick up and move to a more hospitable environment if roads and farms are in the way. And animals can't move outside the parks if the area is developed.

If we let the remnants of wildlife populations remain in isolated reserves, they may become sitting ducks for local extinctions. One method currently being attempted to deal with this problem is to establish corridors between parks and other open lands. Corridors may provide an environmental safeguard should species need to move in the face of global warming. This might require wildlife bridges or tunnels be constructed between open habitats as was done for the Arctic caribou when the Alaskan pipeline was built.

Climatoids

In southern California, biologists use a freeway underpass to connect the Chino Hills State Park and Cleveland National Forest. Corridors like these provide a necessary component to the natural balance of nature. Without the corridor there is insufficient space for larger predators. Mountain lions won't breed and the bobcats would inbreed and eventually die out. There would be an explosion of smaller predators like the fox, skunk, raccoon, and feral cat, which would then put increased pressure on small songbird populations including the California gnat catcher and the least Bell's vireo, both on the Federal Endangered Species List. Corridors like this underpass would also allow wildlife the ability to respond if global warming creates serious challenges to their environment.

The Least You Need to Know

- Global warming is already affecting polar bears near Hudson Bay by breaking up the ice earlier in the year and limiting the bear's access to seals.

- The grizzly bear might be affected if global warming seriously damages white pine in and around Yellowstone National Park.

- Parks, beaches, and other tourist attractions will suffer as the things that draw the tourists are destroyed.

- Global warming will impact sea birds, songbirds, and waterfowl.

- It will do a number on amphibians and reptiles as well.

- National parks might soon have to be connected by freeway underpasses and overpasses if nature is allowed to flee the warming climate.

What's in that Crystal Ball?

In This Chapter

- ◆ The weather 100 years from now
- ◆ How computer models do it
- ◆ What are the variables?
- ◆ How accurate the predictions?

As the inhabitants of this planet, we are engaged in a big experiment. Somewhere around the middle of the twenty-first century, we will have doubled the amount of carbon dioxide in the atmosphere from its preindustrial levels. We don't know exactly when it's going to happen, but it looks pretty inevitable. That means that the level of carbon dioxide, which was 280 parts per million (ppm) in the nineteenth century and 360 ppm in 1995, will rise to 560 ppm. It's the equivalent of dumping several hundred billion tons of carbon in the form of CO_2 into the atmosphere.

We'll be doing some other things as well. We'll be cutting and burning down the rainforest. We'll be plowing up the lands, as well as laying down asphalt and concrete. Nature might even toss in a volcano or two.

Even when we arrive at this landmark goal of 560 ppm, we'll have to wait a while to see the final results. It might take a few decades for the climate to come into equilibrium with this doubled carbon dioxide. It takes a while for the ocean to heat up. In all, it might take a century to fully appraise the damage—if there is any damage.

Global Warnings

By the time we get to the point where we have doubled the carbon dioxide, we'll probably understand the climate a whole lot better. In the process, scientists will have answered a lot of questions about it. We'll know how sunlight, plant life, land surfaces, ice, ocean, and atmosphere affect the climate. We might finally have a good explanation for the ice ages. The experiment may cost us, but at least we'll know better.

Climate Models

Right now scientists are using computers to predict what the answer to our ongoing experiment might be. In many cases they're using some of the same computer programs that predict the weather. Only instead of running these programs out a few days or even a few weeks, they've attempted to run the models out for years, decades, and even centuries.

The models are quite fascinating. Scientists build these models to represent our real world, then they speed them up, turn the sun on and off, pump up the carbon dioxide, and see what happens. These are high-order computer simulations. The rides at Disney World pale in comparison. Even the best of the special effects films have trouble keeping up.

Climate models put our best scientists and our best super computers to the test. There are only about two dozen groups in the world with the right scientists and big enough computers to run these simulations.

Cost Versus Benefits

Running these models isn't cheap. As Richard C. J. Somerville, Professor of Meteorology at the Scripps Institution of Oceanography, describes in his book *The Forgiving Air*, this is big science. He compares the science to astronomical telescopes, particle accelerators, and satellites. Indeed, much of the data that goes into making these long-range predictions is gathered from satellites.

The computers cost around $10 million, which is a lot of money no matter how you look at it. And that doesn't include the building you put the computer in or the staff you use to run it. NASA (National Aeronautics and Space Administration) funds most of it. But weather modeling also shows up as a line item on the ledgers of NOAA (National Oceanic and Atmospheric Administration), the Department of Energy, and the National Science Foundation. NASA leads the way since space research yields much of the data for making these simulations.

Accurate weather reports, the foundation upon which global climate models are built, have tremendous value. Knowing the weather tells a farmer when it is best to plant and when it is best to harvest. But other industries such as shipping, tourism, and various sectors of industry also benefit. Knowing what the weather will be like in the next several days gives these industries a competitive edge. Knowing what the climate will be in the next several years, decades, or centuries could help large industries and whole nations.

The First Weather Report

Lewis Fry Richardson, a mathematician and a physicist who lived from 1881–1953, was the first to try applying mathematics to the problem of weather forecasting. He did much of his groundbreaking work during World War I. Then a devout Quaker and a pacifist, he resolved the conflict between his religious convictions and a desire to see the war up close by driving an ambulance in France. While going back and forth to the front lines to evacuate the wounded, he came up with the arithmetic to make the world's first scientific weather forecast.

Richardson had no super computer. In fact, he did all his math with pen and pencil. He was driven by the idea that he could make the imperfect art of weather prediction into an exact science by reducing the evolution of the atmosphere into mathematical equations. He felt that if we could mathematically predict things like eclipses through calculations, then why couldn't we apply those same principles to the various aspects that make up weather, and predict it as well.

Using calculus, he derived a number of mathematical equations to describe the physical laws that govern such weather phenomena as humidity, pressure, temperature, and wind. His idea was to divide the area of the forecast into a grid or chessboard and gather statistics from the center of each square to build a starting point. He could then factor that information into equations, to determine how things would

Warm Words

Initial conditions is a term that refers to the starting point of a weather forecast that takes into consideration temperature, wind speed and direction, air pressure, and humidity measured from points in a grid overlaying the landscape.

change. The statistics factored all at once for every square were the stating point for the prediction—what scientists called *initial conditions*.

Richardson's first forecast was highly inaccurate. He tried to predict the weather in six hours at a point near Munich, Germany, but was way off. Richardson suspected that his errors were due to poor observation of the initial conditions. In fact, his observing stations were neither numerous enough nor close enough together. He also didn't take into consideration the three-dimensional aspect of the atmosphere. To accurately predict the weather, one has to know not only what the temperature, wind, humidity, and pressure is on the ground, but what it is in the upper atmosphere. Richardson didn't have that option.

Not only were the initial conditions off, but the math was bad. Besides, it took too long. To figure out what the weather was going to be like six hours in advance took him several weeks. By then, obviously, the usefulness of the prediction had passed. But Richardson felt he was on to something. He thought the problems could be worked out. Someday, somebody, somewhere would solve the riddle. His book, *Weather Prediction by Numerical Process*, was published in 1922 and was the first modern treatise on dynamic meteorology. The book described his process in embarrassing detail. But though it was a rough start, it was a start. Richardson was onto something for sure.

Climatoids

One of the biggest problems of Lewis Fry Richardson's first mathematical weather forecast was that it took weeks to do the math in order to calculate the weather for Munich six hours in advance. Though the weather arrived on time, the forecast was incredibly late. Richardson had a remedy. In his book he proposed a large theater filled with thousands of people, each doing the arithmetic for one small part of the calculation. In some ways he was describing the methods used by today's supercomputers. They utilize numerous parallel processors to make the same calculations that Richardson envisioned a theater full of mathematicians for.

The Evolution of Accuracy

Richardson's basic methods evolved. Observations, both on the ground and in the air, got a lot better. The math improved as well. The theories of how the various aspects of weather and climate react got a whole lot closer to the truth. Computers made Richardson's idea of a theater full of people unnecessary.

In the late 1940s a team of meteorologists and mathematicians using one of the earliest computers made the first successful mathematical weather prediction. By the 1950s, meteorologists were routinely making weather forecasts using Richardson's basic methods. Today's climate models are based on those same ideas.

Data In

To predict the weather we must again start out with today's initial conditions. To do that requires a vast amount of measuring instruments, scientists, communications, and a good deal of international cooperation. Measurements are taken at airports, weather stations, and aboard ships. Weather balloons are flown at various stations around the world twice each day to give us accurate readings at higher altitudes. Weather satellites give us a much better picture of cloud formations and the three-dimensional atmosphere.

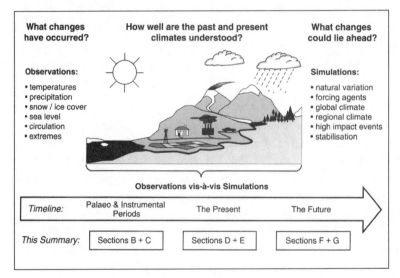

Adding up the data.

(Adapted from Climate Change 2001: The Scientific Basis)

To predict the climate is something else altogether. We are looking for much broader values. We aren't asking if it's going to rain tomorrow, but if it's going to be warmer in the next 10, 50, or 100 years. We aren't asking about the specifics of the weather picture, but overall general aspects. Our grids are farther apart, our information not so detailed, the view one of broad strokes only. We look less at which way the wind is blowing, at what's the temperature in Seattle, but more at larger features like how brightly will the sun shine, how much carbon dioxide will be present, and how much ice will there be.

Hot Debates
Many ask, if we can't predict the weather more than a few weeks out, how can we predict climate 100 years from now? Scientists have had luck predicting El Niños close to a year in advance. El Niño is a climate phenomenon that produces storms in Peru, droughts in Australia, and effects all over the world. Because El Niños last a long time, affect a large area, and involve both the atmosphere and the ocean, they are not an aspect of weather but of climate. In fact, their prediction is our most successful effort at predicting climate to date.

Data Out

Weather services have gotten a lot better with their forecasts. Those from the richest nations utilize computers that equal more than a few theaters full of mathematicians. Not only do we have improved computers but we also have finer resolutions of the grid elements that Richardson used to make his first mathematical weather prediction. The finer the grid, the smaller the squares, the more accurate are the predictions. The typical resolution for today's weather prediction models is about 60 miles (100 kilometers) across each grid element.

Global climate models are much coarser. The grids are spread so far apart that a hurricane or a tornado could pass through unnoticed. Our normal day-to-day weather can't be factored into these models. To run climate models out into the distant future brings most supercomputers to their knees. And that's just doing the broad stroke stuff. The little details like is it going to rain, snow, hail, or hurricane tomorrow, the stuff that's important to us on the ground, won't come out of those models. What comes out is the big picture, most notably increased temperatures. Though it doesn't tell us about hurricanes, it does tell us that the tropical ocean that spawns hurricanes will be warmer. Since hurricanes run on heat, meteorologists extrapolate from that a possible increase in hurricane intensity. Because heat puts more moisture in the air, they also predict that those hurricanes will carry more rain.

Limitations

Weather forecasts have gotten better. We joke about the weather but farmers, air traffic controllers, and shipboard captains know it's not like it was 30 or 40 years ago. Gross errors made in the 1950s and 1960s rarely occur these days. Meteorologists regularly predict weather about three days in advance. Back then you were lucky to get one day. Meteorologists feel that one day we might be able to depict weather as much as three weeks in advance. Much beyond that, however, is unlikely. There is a limit to weather prediction. The reason is that the atmosphere is not a cyclical, periodic phenomenon like the earth's orbit around the sun.

Global Warnings

When the Galveston Hurricane hit that Texas town on September 7, 1900, it arrived totally unexpected. The weather forecast in the Galveston *News* that day was on page two and predicted rain, with high northerly winds, clearing by late Sunday. But by late afternoon Saturday the deadliest hurricane ever to strike the U.S. mainland came ashore and took the lives of over 6,000 people. With today's weather forecasting systems, many of those lives would have been saved.

It's variable, erratic, and random. It's like a pinball machine, no matter how good you are at pinball, when the ball leaves the lever and goes up the track, it hits a peg and bounces around. A small difference in the way the ball hits that peg makes a big difference in how it is going to bounce around. It's the same with golf, tennis, or baseball. A small difference in how the club, racket, or bat hits the ball makes a big difference in where the ball ends up. Mathematicians call this "sensitive dependence on initial conditions."

The atmosphere is like this. It's extremely sensitive to initial conditions. Vary those initial conditions just a little bit and you can end up with a totally different weather outlook. The atmosphere is as unstable as a golf swing. The flap of a butterfly's wings can disturb the initial state and you can end up with a totally different outlook. This was the theory of Edward Lorenz at MIT. He postulated that if you had a planet like earth only it had a butterfly that flapped its wings just once, that flap would eventually result over time in a weather picture very different than the weather on earth.

But there are indications that when we look at the broader-based climate, that this sensitivity to initial conditions may not play as strong a role. Our ability to project an El Niño is an example of a successful attempt to predict long-range weather—the definition of climate. Still, using weather-predicting models to project long-range patterns in climate has an inherent weakness. Mainly, we haven't done it before. We're on new ground. We're out there making new science. And it's untested, untried, and has major consequences.

Feedback Mechanisms

Meteorologists believe that in making long-range climate predictions, the crux of the problem is determining to what extent various feedback mechanisms will affect the climate. When global climate models are run on our experiment of a doubling of the CO_2, the models predict on average worldwide increase in temperature of from 1.8–7.2°F (1–4°C). The variation is primarily due to how the models handle feedback mechanisms such as ice, soil, ocean, water vapor, and clouds.

The difficult part of climate modeling is predicting how those feedback mechanisms will work. If there weren't any feedback mechanisms, we could figure out the temperature increase on Earth from a doubling of carbon dioxide by simple math. The answer is that the world would increase its temperature by about 1.8°F (1°C). The models, however, show that the Arctic temperatures could rise 21.6°F (12°C) and that's because tremendous feedback mechanisms come into play in the Arctic.

Ice

According to global climate models, if you double the amount of carbon dioxide in the atmosphere, the greatest amount of warming will be in the far north. The reason is that in the Arctic there are vast expanses of ice and snow. And as the atmosphere warms up, the ice will start to melt. Already 40 percent of the Arctic ice has melted. As the temperature gets warmer, the melting will accelerate. By the time the world has reached equilibrium from the doubling of carbon dioxide, half, three quarters, or even all of the ice will have melted in some areas.

Darker ocean, bare land, and grass-covered land will replace the ice. These are all darker than ice or snow and have lower albedo, lower reflectivity. So instead of the sun's radiation being reflected back, it will be absorbed by the darker surfaces. Anyone who's walked barefoot to the beach in the summer knows that the darker asphalt road is a lot hotter to tread upon than the lighter concrete sidewalk. That's because the dark asphalt absorbs all the heat, while the lighter concrete reflects some of it and in the process stays cooler.

> **Climatoids**
>
> The amplification of the global warming that computer models predict for the northern latitudes is caused by the ice-albedo effect in which melting ice leaves surfaces that are more absorbent of heat. The effect is more pronounced in the high northern latitudes than in the high southern latitudes because the southern hemisphere has less land to be covered with snow and ice.

The computers predict that the greatest warming will be in the north, because there is a positive feedback system. As the ice melts, the darker water or earth beneath it absorbs more heat. The change in the ice due to warming amplifies the warming. Estimating how much of an effect this will have accounts for some of the variation in climate-model predictions.

As the temperature gets warmer, the melting will accelerate.

(Photo by Michael Tennesen)

Soil Moisture

And then we have the tropics where there isn't any ice. But there are feedback mechanisms here as well. One of the problems modelers have in the tropics is determining the amount of feedback in soil moisture. The reason is that when the sun's rays hit the surface of the earth in a place where the soil is moist, part of the energy of the incoming rays is used to evaporate the water. If the soil is dry, then all the energy in the earth's rays goes to heat the soil, rather than evaporate the water. If the sun has to tackle soil moisture, that's going to take some of its heat away. If the soil is dry, then all its energy goes into the air, making it hotter. This is also one of the reasons that the surface temperatures in the desert can get hotter than the tropics, though the tropics are closer to the equator and thus receive more direct light from the sun. Desert soils are dry and absorb more heat.

Rain and Snow

Scientists estimate that if nothing is done about the rising levels of greenhouse gases, then the oceans will rise about 20 inches (50 centimeters) by the year 2100. The reason is that as the ocean becomes warmer, it expands—warmer water takes up more room than colder water—and the ice that is on the land melts and flows into the sea. During the ice age when more of the water was contained in glaciers, sea level was much lower than it is now. There was a land bridge that connected Alaska and Siberia.

But one of the things that happens with global warming is that as the temperatures rise and the ocean warms, more of the ocean evaporates into the atmosphere and it rains and snows more. As a result, Antarctica and Greenland will lose ice into the sea because of increased temperatures and increased melting, but there will be more snow, so the amount of polar ice could actually increase.

Water Vapor

Climate models have long included a positive feedback mechanism from water vapor. Warm air holds more water than cold air. Air in the tropics is more humid than air at higher latitudes. As the temperature warms up, so will the ocean. A warmer ocean will evaporate more water into the atmosphere. And this additional water vapor increases the greenhouse effect because water itself is a greenhouse gas, more powerful than carbon dioxide.

In many ways, carbon dioxide is simply the trigger by which more water vapor enters the atmosphere. This water vapor will amplify the warming caused by the carbon dioxide. Because of this, the temperature rise from an increase in carbon dioxide might be 1.5 to 3 times as large as it would be without this feedback. Climate models have to deal with all of these interrelated phenomena in order to predict the future.

Clouds

Clouds may be the trickiest feedback mechanism of all. They have a dual role. First, they reflect sunlight coming in from the sun, and redirect it back into space. Clouds reflect about 20 percent of the sunlight directed at Earth back into space. The effects vary with cloud height and type. Since less sunlight falls on the earth, less heat is absorbed, and so it's cooler. Thus clouds by their reflectivity cool the earth.

But, wouldn't you know it, they also warm it up. The clouds absorb the rays of the sun that bounce off the surface of the earth, and they redirect that heat back toward Earth, which increases the temperature. It's the greenhouse effect again. Remember the glass over the greenhouse kept the plants beneath it warmer and moister. Well, that's just what carbon dioxide, methane, water vapor, *and* clouds do. They keep the earth beneath it warmer and moister.

Factoring in the clouds.

(Photo by Michael Tennesen)

So which dominates? Satellites have been deployed to determine the contribution to the greenhouse effect and to Earth reflectivity from clouds. According to Richard Somerville, who has spent a good deal of his meteorological career studying the feedback effect of clouds, on average clouds cool the planet. In other words, the reflectivity of clouds dominates over their greenhouse effect. He translates the measure of their strength into watts, the same watts that power your light bulbs. In general, clouds reflect away an average of 45 watts of sunlight per square meter of the earth's service. The greenhouse effect of clouds contributes about 30 watts of heat per square meter.

But that's without global warming. When you add the warming, cloud cover might decrease. Clouds currently cover approximately 50 percent of the surface of the earth. In a modeled double–carbon dioxide atmosphere, cloud cover is decreased a few percent. Since the cooling effect of clouds dominates over the warming effect, when you remove clouds you are removing more cooling than warming. The net effect is it might get warmer. It's similar to removing the ice. The surface beneath is darker and absorbs more heat. Additionally in a carbon dioxide–doubled world, the clouds on average are higher. And the higher the clouds go, the colder they get. Colder clouds don't radiate as much heat. According to Somerville, the difference in the degree of warming predicted by the models is principally due to how the models handle the difficult question of cloud feedback.

Climatoids

The earth reflects about 30 percent of the sunlight that lands on its surface. Two thirds of that is due to the reflectivity of clouds and one third is due to the reflectivity of ice, snow, and the bright surfaces of the desert.

The Forecast

All the climate models predict that the earth is going to get warmer. By the time the carbon dioxide rises to 560 ppm, double its preindustrial levels, the atmosphere could warm approximately 1.8–7.2°F (1–4°C). The IPCC predicts even greater warming by 2100 when CO_2 could rise from 540 to 970 ppm. Though there is a range of degrees, no one is expecting it to cool down. Taking a temperature of the earth is a little like taking the temperature of a sick patient. Is a 1°F (.6°C) rise in temperature enough to worry about? Maybe, maybe not. But what about 2°F, 3°F, or 4°F (1.2°C, 1.8°C, or 2.4°C)? That could indicate a serious underlying infection. And what about a 7.2°F (4°C) rise in temperature? That could indicate a fatal illness.

Should We Trust Them?

The trouble is we've never done this before. Still, the best efforts of about 2,500 of the world's leading climatologists and meteorologists shouldn't be taken for granted, though even they have doubts themselves. But in 50 to 100 years, if their predictions prove wrong, we won't have lost much. We might even have a little gas left. We might have some new technologies that don't pollute the air as much. There might be some rainforests left. But in 50 years, if we've done nothing to turn the tide of increased greenhouse gases, it might be too late. That's because at whatever point we decided to stop pouring carbon dioxide into the atmosphere, the results we're looking at then might just have to be lived with for the next 100 years, since that's how long carbon dioxide persists in the air.

The Least You Need to Know

- Global-climate models run on the same machines that predict the weather.

- The difference is that weather prediction models try to predict the detail, while climate models look for big-ticket items only.

- Weather models and climate models have lots of problems, but the models are still very useful.

- Snow, ice, water vapor, cloud cover, and other feedback mechanisms affect climate-model predictions.

- All models predict warming, but the range of temperature increase is broad.

Part **5**

The Politics

In the late 1980s, the nations of the world came together to tackle the problem of ozone depletion. Certain gases, chlorofluorocarbons (CFCs), were creating a hole in the ozone layer over Antarctica. The national community gradually outlawed CFCs. It was a great moment. We had all gotten together and done something for the atmosphere. How come we can't do anything about global warming?

I'll look at the Kyoto treaty. What was it all about, anyway? How did this credit system work—trading smokestacks for rainforests? How does U.S. politics fit into this? I'll look at the international debate—who's skeptical and who's not. Could pollution reduction, energy conservation, and rainforest preservation be worthwhile endeavors all by themselves?

Finally, I'll sum up the argument for you. Who's behind the predictions, what are they saying, and what's the current evidence? Should we move forward, or are the costs too expensive?

Chapter 18

A Treaty Known as Kyoto

In This Chapter

- ◆ Bartering with carbon dioxide
- ◆ Cut pollution or grow trees
- ◆ Developing countries let off the hook
- ◆ The United States backs out

In December 1997, over 160 nations met at the United Nations Conference on Climate Change in Kyoto, Japan, to develop an international treaty to deal with the growing body of evidence that global warming was a reality and the world had to do something about it. At the Earth Summit in Rio in 1992, the Parties to the Framework Convention on Climate Change were given the task of coming up with an agreement that would halt the increasing flow of man-made greenhouse gases into the atmosphere. Five years later, the parties had come together in Kyoto to try and work out an agreement. It wasn't an easy task.

The Signing

After 10 days of marathon meetings and late-night negotiations, agreement was reached, and they finally put pen to paper. In the end, after 160 nations had wrangled, fought, and played tug-of-war, they passed an important milestone by acknowledging a global threat and coming up with a global solution to protect the earth's environment.

This was different from Montreal when in 1987 representatives from all over the world had come together to ban CFCs. Carbon dioxide, methane, and other greenhouse gases required a far grander and more monumental effort. This was a lot more than just swapping out the gas in the refrigerator. To control greenhouse gases required whole new technologies, whole new sets of tools and machinery, indeed a whole new lifestyle.

We created carbon dioxide with automobiles, industry, cooking fires, plows, axes—even our breath. We had to take a hard look at the cost of modern transportation. What price had the atmosphere paid for the increasing efficiency with which we moved our products to the market? What were the real costs of cutting down the rainforests in terms of climate change?

The cost of modern transportation.

(Photo by Michael Tennesen)

It seemed now that the world might finally be ready to do something about it. Though the scope of the agreement was limited, its mechanisms were important because it set standards for further agreements, and created tactics from which the battle could be fought and hopefully won. It recognized that concentrations of certain greenhouse gases had been growing since the beginning of the Industrial Revolution. And it recognized the international importance of reversing the steady increases.

Divvying Up the Burden

The idea was that 38 nations needed to reduce their greenhouse gas emissions by an average of 5.2 percent below 1990 levels. In reality, the reductions were parceled out according to who emitted the most gases and who could afford the changeover. The United States would reduce emissions by 7 percent below 1990 levels, Japan by 6 percent, and the European Union by 8 percent. The Russian Federation, Ukraine, and New Zealand needed to return to 1990 levels, while Australia and Iceland had to stabilize their releases at 8 percent and 10 percent, respectively, above 1990 levels.

The European Union wanted to get started right away, but the United States pressed to have enforcement of the reductions put off until 2008. This would provide a full decade to allow governments and industries to gradually shift emphasis toward increasing energy efficiency: to upgrade pollution equipment as older machinery wore out, to switch from dirty fuels to cleaner-burning energy, and to experiment with solar, wind, and other alternative forms of energy.

Getting 160 nations to agree on something wasn't an easy task. European Union delegates had arrived with the ambitious proposal for all industrial nations to cut their emissions by 15 percent below 1990 levels. It seemed easy for them to talk. The collapse of the former East German industries and a massive switchover in Great Britain from coal to natural gas during the 1990s had given Europe a head start on the rest of the world. But Europe relied heavily on emission-free nuclear power, and nuclear power wasn't exactly squeaky clean.

> **Hot Debate**
>
> History played a vital part in the Kyoto negotiations. One of the reasons that industrialized nations were asked to go first in the negotiations was because of the longevity of carbon dioxide in the atmosphere. Much of the carbon dioxide that was released in the Victorian Era by those nations was still in the air. Developing nations had only just begun to add to the problem.

> **Hot Debate**
>
> In trying to reach greenhouse gas emission standards, many at Kyoto complained that the European reliance on nuclear power for *clean* emissions wasn't all that clean. Even hydroelectric dams in use by the United States and many developing nations were considered by many to be dirty. Hydroelectric energy was often created at a high cost to wildlife and the natural environment.

But the real struggle was in the details. Japan insisted it wouldn't go any higher than 5 percent in emissions reductions. A 2 A.M. phone call from then vice president Al Gore to Prime Minister Ryutaro Hashimoto got Japan to up its ante. Gore argued that Japan couldn't wreck the chances to come to an agreement over a measly 1 percent. The Japanese prime minister relented.

There were other difficult points. At Kyoto, Russia and Ukraine were given the right to emit as much greenhouse gas in 2010 as they did in 1990, the year after the Berlin Wall came down. But their industries had collapsed and their emissions were as much as 30 percent below 1990 levels already. Also, while the new treaty would impose binding limits on emissions by 39 nations, 121 less economically developed nations escaped any kind of reduction, even though some of them were expected to grow rapidly. It totally let both China and India off the hook. This was in spite of the fact that their emissions were growing rapidly and were eventually expected to lead the world.

Climatoid

One of the exemptions to the Kyoto Protocol was emissions from the military. The United States pushed for and won exemptions that were carried out under the United Nations charter. The idea was that nations might hold back from participating in humanitarian, peacekeeping, and other operations because of concerns about greenhouse gases.

The agreement placed limits on six greenhouse gases: carbon dioxide, methane, nitrous oxide, hydrofluorocarbons, perfluorocarbons, and sulfur hexafluoride. The European Union had wanted to limit the new curbs to just carbon dioxide, methane, and nitrous oxide. But the United States succeeded in adding limits on hydrocarbons, perfluorocarbons, and sulfur hexafluoride. The later chemicals were arguably easier and cheaper to cut.

Is It Enough?

Would the Kyoto Protocol solve the problem of global warming? Hardly. According to the IPCC, the convention's scientific advisers, the treaty, if passed, would buy 10 more years at most. To halt global warming, they said, would take a much more radical commitment. The IPCC suggested that the world should agree to a maximum concentration of greenhouse gases in the atmosphere. A reasonable goal might be a doubling of carbon dioxide in the atmosphere over preindustrial levels. That figure represented an increase of about 50 percent above then present-day levels.

Many had already argued that a doubling was dangerous. But scientists were ready to acknowledge that that doubling was likely to take place, treaty or no. But at least if it didn't get any worse, that would be something. Although the Kyoto agreements did not call for the levels of emission cuts most scientists considered necessary, it was a start. It was a first step in the right direction, an acknowledgement of the problem and a resolve to find an answer.

Bartering

Despite its varied drawbacks, the Kyoto agreement did contain one breakthrough that many felt was significant. It allowed industrial nations to reduce emissions through a credit-trading system. The concept had been pioneered by George H.W. Bush in the Clean Air Act, which in the early 1990s cut acid rain in half by allowing U.S. utilities to trade sulfur dioxide credits. A polluting industry could buy or trade credits with another industry that had exceeded its pollution-control goals.

A similar idea was proposed at the Kyoto conference. The new system would permit the industrialized nations to trade carbon-emission credits amongst one another and with developing countries to reach their own greenhouse gas emission goals. Since global warming was an international problem, setting up a global trading regime seemed like a good idea.

A country with the potential for wind power or other source of renewable energy might develop them on a large scale and potentially reduce their greenhouse gas emissions to below required levels. They could sell the credits for the extra reductions to countries which didn't have as great a potential for renewables. It might be more cost-effective for an industrialized nation—particularly one who had already gone through significant pollution reduction—to simply buy the credits of another nation rather than to go through more expensive reduction methods.

The fear was, however, that some countries might find themselves with spare credits to sell, just because their economies were in the dumps. Such was the case with Russia and Ukraine. The Kyoto agreement gave them pollution limits that were quite a bit above what they were producing in their then current economic downturn. Although Russia promised to invest any of the monies it might gain in greenhouse gas credit sales into cleaner technologies, some people saw this type of trading as a loophole in the basic role of global-emissions reductions. Environmentalists referred to it as trading for *hot air*, a reference to the belief that this type of trade would yield little in the way of the conference goal of making concrete efforts to reduce greenhouse gas emissions.

Warm Words

Hot air is a term used by environmentalists to apply to greenhouse-gas reductions that resulted from weakened economies and the resultant reduction in manufacturing rather than actual efforts put into making those industries operate cleaner and more efficiently.

Cleaning Up Someone Else's Backyard

In some cases it was cheaper and easier to clean up someone else's mess rather than attend to stuff at home that has a higher price tag on it. Kyoto allowed for this. Under the agreement, a country could invest or in other ways participate in an emissions reduction program in another industrialized nation, and gain credits that could be applied toward the emission reduction goals at home.

Climatoid

One of the basic beliefs behind the global emissions trading under the Kyoto agreement was that it didn't matter where one chose to lower emissions, whether at home or abroad. A reduction of the emissions on one side of the globe will eventually affect the other side of the globe in equal portions, principally due to the fact that the gases are so long lived and eventually mix in equal portions over the entire world.

The treaty also allowed wealthy nations to invest in emissions reduction projects in poorer nations and get credit as well. Not only might it be cheaper to invest in emissions reduction in the third world, but that investment could become a major engine for getting clean air technologies into poorer nations. This could prevent those nations from going down the same polluted path that industrialized nations previously navigated and now have to amend.

What About Methane?

The emphasis in the Kyoto agreement was on carbon dioxide. Of all the greenhouse gases, it is the longest lived, lasting up to 100 years in the atmosphere. It is also responsible for the largest portion of the greenhouse effect.

Climatoid

Methane is the second-most important greenhouse gas, yet it is much more potent. Releasing 22 pounds (10 kilograms) of methane into the air today will warm the world about as much over the next decade as a ton (907 kilograms) of carbon dioxide.

Some scientists, however, felt that more attention should be given to methane. Methane has a much stronger greenhouse effect than carbon dioxide. The reason that carbon dioxide was given greater importance in the agreement was that methane only lasted a decade in the atmosphere. But a little bit of methane was equal to a whole lot of carbon dioxide in terms of climate change, so that by decreasing methane there was a potential for a more immediate reversal of global-warming trends.

Plus, methane is easier to clean up in the atmosphere. All you have to do is fix the natural gas pipes so they don't leak so much. Put a little more soil or inert waste on landfills to encourage the growth of methane-eating bacteria. Maybe feed the cows something that doesn't give them so much gas.

In addition, cost-effective methods, such as plugging leaks in gas lines, would lead to savings over time, a counterpoint to those who argue that reducing greenhouse gases would hurt their economies.

Two Smokestacks for One Rainforest

One of the methods for gaining emissions credits envisioned by the Kyoto agreement was to encourage the land to soak up more carbon dioxide. That meant creating man-made *carbon sinks* by planting trees or other vegetation, or by practicing various types of soil conservation.

The treaty allowed industrial nations to receive some emissions credit for any land it had forested since 1990. However, they also had to increase their emissions reductions if they had deforested any of their land since that date. So one could receive credit for creating a carbon sink, but would also get penalized for destroying a carbon sink.

> **Warm Words**
>
> **Carbon sinks** are areas like forests, grasslands, and the oceans, which naturally take up carbon dioxide from the atmosphere.

Countries could also claim credit for planting forests in developing nations. The draw here was that an industrialized nation might plant a forest in the tropics, for example, where trees would grow a lot faster and land wouldn't be as costly. Not only would this decrease the amount of greenhouse gases in the air, but it would decrease tropical deforestation, which had other benefits, such as increasing habitat for wildlife.

Shallow Plow

Countries could also claim credit for soil conservation measures, which would allow more carbon to be soaked up by the soil. One idea was *low-till farming*, a method of turning the ground that could trap more carbon in the soil. Deep plowing allows more oxygen into the soil, and speeds up the decay of organic matter and the release of carbon dioxide into the atmosphere. Set the plow a little shallower and it would decrease those emissions. Countries like Canada and Australia professed that by such soil conservation methods they could soak up tens of millions of tons of carbon each year.

Factoring in the rainforest.

(Photo by Michael Tennesen)

The European Union came into these talks feeling that these types of mitigation would only be a small part of the efforts to meet emissions targets, whereas the United States, Russia, Canada, and some other countries thought they should be a major part of the treaty. The traders eventually won out in the drafting of the agreement.

The Price of Carbon Dioxide

Even though bartering seemed like a good idea, problems arose from trying to figure out how much to charge for emissions. In other words, how much is a pound or a ton of carbon dioxide worth? How does one accurately determine the value of a carbon sink? How does the world patrol individual country emissions? Will there be an international greenhouse police, or will each country monitor itself? Also, how will emissions trades be made? Will we print emissions credits? What bank will handle the transactions?

The problems of organizing and maintaining such an international monitoring and trade organization were difficult to say the least, but countries like the United States had had success with trading sulfur dioxide emissions. Smaller municipalities like

Los Angeles had allowed businesses to achieve credits for pollution reduction, not by adding new technologies to their smoke stacks, but by buying up older more polluting cars off the streets and sending them to the junk yard.

Your Bill, Sir

What all this would cost the economy was a point of contention. When Kathleen McGinty, chair, Council on Environmental Quality for the Clinton administration, went before the U.S. Congress to report on the cost of the proposed treaty, she estimated that the cost of the treaty to the average family's energy bill would be about $70 to $110 a year in 2010. Industry estimates have been much higher.

Furthermore, McGinty noted that economists had estimated that the environmental, health, and economic costs of global warming projected to occur from a doubling of carbon dioxide in the atmosphere were about $80 billion a year. One of the secondary benefits of greenhouse gas reductions was a decrease in pollution. McGinty estimated that this reduction in pollution alone could result in a saving in health care costs equivalent to one fourth of what the United States would pay to achieve its greenhouse emission-reduction target.

Hot Debate

Though the Clinton administration originally proposed that the increase in the average family's energy bill would be about $70 to $100 *annually* in 2010, others have put the price far higher. A study funded by the petroleum industry concluded that meeting greenhouse gas emissions targets in the Kyoto treaty could raise the family's *monthly* expenses in natural gas by $92, heating oil by $110, electricity by $48, and gasoline by $53. That amounts to a total *annual* increase, according to the petroleum industry, of $3,636 compared to the Clinton administration estimate of $70 to $110.

Those in the U.S. petroleum industry were not the only ones concerned with the costs of implementing the Kyoto agreement. Though the Canadian government had pressed toward ratification of the treaty, Canadian industries, particularly in Ottawa, resisted. One report by the Canadian Manufacturers and Exporters Association claimed that 450,000 jobs would be lost because of Kyoto. The government countered that only about 61,000 would go, or, under a worst-case scenario, as many as 244,000.

Developing Countries

Under the Kyoto treaty, the world's developing nations were not required to make reductions in greenhouse gases, although Argentina and Kazakhstan nevertheless agreed to do so. The reasoning behind this was that industrialized nations, the United States in particular, were responsible for a disproportionate amount of greenhouse gases. The United States, though it had only 4 percent of the world's population, was responsible for 25 percent of the world's greenhouse emissions.

Since most of the carbon dioxide in the atmosphere to date was the responsibility of the industrialized nations, it was agreed that they would take the first step. Then, perhaps, the developing countries would be more willing and eventually more able to follow suit. Poor countries insisted that the excesses of the rich nations were largely responsible for the problem, and that cutting emissions would come at the price of economic growth, making their lives even poorer.

Yet it is the developing countries which will be producing the lion's share of global greenhouse emissions by the end of the twenty-first century. Argentina, at a follow-up conference on global warming in Buenos Aires, attempted to break the impasse by offering to make voluntary cuts and encouraging others to do the same. Countries such as China and India refused to debate the matter, yet China was expected to lead the world in the production of greenhouse gases by the end of the twenty-first century.

Still, more than 80 countries signed the Kyoto agreement, though the treaty has not gone into effect, because not enough of these countries have ratified it. The United States has signed, but not ratified the agreement, which required a two-thirds majority in Congress to take effect.

The U.S. Balks

Many in the U.S. Congress couldn't handle the lack of participation by the developing nations. A resolution in the U.S. Senate in 1997 which passed 95–0 put the Senate starkly against ratifying a treaty that required no meaningful participation by developing nations. The United States demanded that developing nations should accept their own specific emissions targets—even if those targets allowed emissions to increase.

But the biggest setback for the treaty came in 2001 when the Bush administration announced that it was explicitly opposed to the Kyoto agreement. Environmentalists from around the world greeted the announcement with loud dismay. Despite the growing evidence of global warming and the undeniable role of the United States in creating it, the administration had passed on the challenge.

Climatoid _____

Despite the fact that the United States has pulled out of the Kyoto agreement, many businesses have come to recognize global warming as a threat and the goal of reducing emissions as being in their economic interest. Several such industries such as BP-Amoco, IBM, and Johnson & Johnson have voluntarily adopted emission-reduction goals that are stronger than those required by Kyoto.

The Rest of the World Forges Ahead

Some thought that when the United States backed out of the Kyoto agreement, that this would spell death to the treaty. But in some ways, it has proved just the opposite. Japan and the 15 countries of the European Union have ratified the agreement. Canada, though they had had problems in that country with the agreement, ratified it in December 2002. It seems that by pulling out of the agreement, the United States might actually have helped to unite the rest of the world, which appears now more determined than ever to sign the agreement even if the biggest polluter is not along for the ride.

The protocol needs to be ratified by nations that together account for 55 percent of global greenhouse gas emissions before it takes effect. At press time 116 nations constituting 44 percent of the emissions have ratified it. The treaty has been formally ratified by all the current members of the European Union. They've asked the United States to end its isolation and sign on.

Australia said it will not ratify the Kyoto pact on global warming unless the United States and developing countries get fully involved. Many countries including Australia feel strongly that the protocol will not be meaningful without the participation of the United States.

In the end, if the Kyoto treaty is ratified, the United States will have to deal with a world run under the agreement. Businesses that want to engage in international trade will have to abide by its rules if they want to trade outside of the country. U.S. manufacturers will have to adhere to its principles if they want to sell their goods abroad.

If the rest of the world moves forward on this agreement, can the United States be that far behind?

The Least You Need to Know

♦ The Kyoto agreement is the closest the world has come to date on a plan to fight global warming.

♦ The agreement divvied up the burden of reducing greenhouse gases amongst the wealthiest industrialized nations.

♦ Third-world nations were let off the hook for the time being, a point of contention in the U.S. Congress.

♦ The agreement allowed a system of barter whereby polluting nations could buy credits from nations whose reductions exceeded their requirements.

♦ Nations could gain credits by reducing greenhouse gases in other countries or by planting forests, which take up carbon dioxide.

♦ The pullout of the United States from the treaty has not been the end of the road for Kyoto that many thought it would be.

Chapter 19

What If It's a Bunch of Bull?

In This Chapter

- ◆ A critical debate
- ◆ The international controversy
- ◆ U.S. politics
- ◆ The merits of pollution reduction, energy conservation, and rainforest preservation

The debate over the Kyoto agreement encompasses the basic premises and core arguments of what to do about global warming. Despite the evidence, many still doubt whether global warming is a reality. The evidence, though it looks good in all those charts, still amounts to 1.1°F (.6°C) of warmth. The price tag for the cure is nothing to sneer at. Who's going to pick up the bill? How do we regulate trades? How much of this is politics?

Pros and Cons

Few modern issues are as confusing to the average citizen or elicit as much controversy as the arguments over global warming. People with no background in science or weather are not afraid to weigh in on a discussion of its impact and legitimacy. Many of the discussions are heated. Scientific meetings on the topic have been known to take on the rancor of a bar room full of sports fans arguing over their favorite teams. "I didn't find that in my model," is a common refrain at international gatherings. So let's take a look at some of the arguments. Both sides have legitimate points.

Is Kyoto the Answer?

Proponents argue that if we are going to deal with greenhouse gases, it has to be on a global scale. Individual efforts haven't worked. It took an international effort to address ozone depletion; it will take a similar global effort to tackle global warming. Though there may be flaws in the Kyoto treaty, right now it's the only game in town. Though its scope is minimal, at least it's a start.

Detractors believe that we may need an international approach, but it should include developing areas like China, India, Mexico, and Latin America where emissions are rapidly rising. These countries are not now subject to Kyoto emission-reduction targets. Also, the United States and Australia need to participate, otherwise the rest of the world is shouldering an unfair burden.

Climatoids

1998 Carbon Dioxide Emissions Per Capita in Tons:

United States	20	South Korea	8
Canada	17	Ukraine	7
Australia	16	France	7
Netherlands	15	Poland	7
Saudi Arabia	12	Spain	7
Russia	10	Iran	5
Germany	10	Turkey	3
Taiwan	10	China	2
United Kingdom	9	Brazil	2
Japan	8	India	1
South Africa	8	Indonesia	1
Italy	8		

Source: U.S. Environmental Protection Agency

Is It Worth the Effort?

One side admits some scientific uncertainty about the extent of global warming and its consequences. Still, it seems prudent to take some preventative measures now, since the buildup of carbon dioxide in the atmosphere is cumulative. If we wait until its effects become more evident, we may have to live with those effects for a century or more.

The other side professes that though carbon emissions may be significantly reduced at relatively reasonable costs, entirely stopping climate change will require 50 to 60 percent cuts. This is likely to be hugely expensive, requiring major reductions in the use of gasoline, fuel oil, natural gas, and coal.

Do we need all that oil?

(Photo by Michael Tennesen)

What About Voluntary Measures?

Proponents of voluntary measures argue that business must make emissions reductions gradually. The scope of the changes is not something that can be done overnight. Technologies need to be developed, capital investments planned far in advance, and markets nurtured. Businesses need the flexibility to make those changes over time.

Detractors state that the United States made a commitment to return the nation's emissions to 1990 levels, but in the last five years there has been a 4.6 percent emissions growth. Canada made a commitment in 1990 to stabilize its

> **CAUTION**
>
> ### Global Warnings
>
> The U.S. Department of Energy announced in 2002 that the United States increased its annual greenhouse gas emissions from 5,397 million tons of carbon dioxide in 1996 to 5,647 million tons in 2001 due to the nation's reliance on coal, oil, and natural gas. That number represents a 4.6 percent increase over five years.

greenhouse gas emissions at 1990 levels by the year 2,000. But come 2,000, they were actually 20 percent above those levels.

Is the Technology in Place?

Proponents of Kyoto argue that we can go a long way toward meeting our 2010 target with existing technologies. But to meet the longer-term goals, we are going to need new technologies. To get there, we need to start making the investments now.

Detractors state that the only way we can meet the Kyoto goals is by accelerating technological development so that we can avail ourselves of new technologies and new fuels. These technologies could be in place within a decade, but that won't be fast enough. To expect that amount of progress within the time frame of the Kyoto agreement is unrealistic.

Is It Too Great a Burden?

There have been a number of cost benefit analyses that have tried to answer the question of whether the human and environmental benefits from eliminating or slowing global warming exceed the costs of cutting emissions. The World Resources Institute reviewed 16 of these simulation studies and concluded that reducing carbon emissions to 1990 levels by 2010 would affect the U.S. Gross Domestic Product (GDP) by a range of a 2.4 percent loss to a 2.4 percent gain. Making the recommended reductions of 50 to 60 percent could cost as much as 7 percent of the GDP, which is $530 billion, or $2,000 for every man, woman, and child in the country. Seven percent of the GDP is twice the annual military budget, three times the federal cost of Medicare, and one and a half times the cost of Social Security. That amount of money could go a long way toward ending social problems such as poverty and unemployment.

On the other side, a coalition of environmental groups, including the Union of Concerned Scientists and the Natural Resources Defense Council, published their own analysis, Energy Innovations, which argues that North America has an enormous potential for improving energy efficiency and developing renewable sources such as solar and wind power. They argue that reducing carbon dioxide emissions by 30 to 40 percent can be achieved while saving money—primarily because implementing energy conservation measures is cheaper than using energy wastefully. The National Wildlife Federation states that many of these studies fail to mention the potentially enormous costs of inaction. Economic estimates reach as high as $400 billion a year by the middle of the next century if greenhouse gas emissions continue at the current pace. Those costs include losses to coastal property due to rising sea levels, agriculture due to droughts and floods, and infrastructure due to increased hurricanes and other major storms.

> **Climatoids** _____
>
> STMicroelectronics, a Geneva-based $7.8 billion semiconductor manufacturer, believes that protection of the environment is not incompatible with economic growth. In the 1990s, they began committing 2 percent of their annual capital investment to improving their environmental performance. By 2001 they were using 28 percent less electricity than they did in 1994. They've committed to zero-equivalent carbon emissions by 2010. To do this they plan on using from 5 to 10 percent renewable sources such as wind and solar power. They will then investigate other reductions. Finally, they plan on planting enough trees to make up for the emissions they still produce. President and CEO Pasquale Pistorio says that by the time they reach their goal, they'll save $900 million a year in reduced energy bills.

Is It Healthy?

The IPPC predicts that global warming could bring increased floods, droughts, and forest fires to varying places. We have only to look at the tragedies caused by the 1997–1998 El Niño to understand some of the tragic potential that may be in store for us if the weather turns extreme. Disease will spread as ticks and mosquitoes move poleward from the tropics, carrying deadly infections.

Detractors say that's all hypothetical. What Kyoto will most definitely do is increase fuel bills. Higher fuel bills will price air conditioning out of the reach of poorer consumers. While this might not harm young people, older people are more susceptible. In the 1995 Chicago heat wave, some 339 seniors perished from heat exhaustion. Higher fuel costs might turn some to wood-burning fires, which can cause respiratory problems for children.

Then there's the question of mandatory increases in fuel efficiency for cars, a lot of which will come from building smaller, lighter vehicles. Despite air bags and tougher materials, collisions in smaller cars usually mean trouble. In 1991, the National Highway Traffic Safety Administration concluded that the federally mandated drop in vehicle weights—from 3,700 to 2,700 pounds in the 1970s—boosted highway fatalities by 2,000 annually and serious injuries by 20,000.

On the other hand: Recent studies on accidents and SUVs revealed that SUVs are not as safe as previously assumed. Accidents in SUVs have a high percentage of fatalities due to their tendency to roll over.

> ### Hot Debates
>
> The IPCC's *Climate Change 2001* assessed the global price of stabilizing atmospheric carbon dioxide at twice preindustrial levels by 2100 at around $1 trillion to $8 trillion. Though that sounds like a lot, Steven Schneider, professor of biological sciences at Stanford University, claims it's really minimal. Economists estimate the world will be 10 times richer by 2100. When you add on the costs of tackling global warming, it postpones that target by two years. That means that the world will have to put off being 10 times richer until 2102 instead of 2100. Similarly meeting the standards set down by the Kyoto agreement means that industrialized nations get 20 percent richer by June 2010 rather than by January 2010. According to Schneider, nobody will know the difference.

Is the Science Sound?

Anti-Kyoto petitions have flooded the boardrooms, union halls, and the nation's capitals. One petition known as the "Petition Project" was signed by 15,000 scientists of various specializations, but with no official sponsor, urging the U.S. government to reject the Kyoto treaty based on a lack of convincing evidence. At congressional hearings the testimony of a few climate skeptics, many funded by industry, created the impression of considerable climate uncertainty.

> ### Hot Debates
>
> Businesses say that to increase the cost of doing business in North America without tax incentives is simply going to move production to developing nations that don't have to adhere to Kyoto or greenhouse gas emissions reductions. Since scientists tell us these gases mix globally, we might lose the business but the global warming produced elsewhere will come back to haunt us.

On the other hand, the IPCC, not a group of unrelated scientists or industry spokespersons but an expert committee of the world's leading climatologists, has spent over a decade conducting comprehensive, peer-reviewed assessments of climate change. Their conclusion is that humans are, indeed, influencing the global climate and that climate change is underway.

The International Debate

The international debate takes on different shades of opinion. Countries in the southern hemisphere feel that the problem is one that was created by countries in the northern hemisphere, but which will affect them greater since disease, inclement weather, and environmental effects are expected to hit the southern hemisphere the hardest. Without financial resources, they won't be able to adapt as well, either. The international picture is certainly far from homogenized.

Developing Countries

Members of OPEC, particularly Saudi Arabia and Kuwait, have been strong opponents of Kyoto. After all, they make their living off fossil fuels. Much of the developing world, which already accounts for around 40 percent of the greenhouse gas emissions, has also been against the agreement.

They feel the rich nations are largely responsible for the problem, and that the current level of greenhouse gases in the atmosphere is the product of 150 years of development by those nations. Now that they have theirs, the rich nations want to put the brakes on the growth in the developing world.

Developing nations argue that they need cheap energy to fuel growth. China, with one third of the world's proven reserves of coal, relies on coal for about 80 percent of its energy needs. It plans on expanding coal production fivefold by 2020. Brazil and Malaysia, home to a large percentage of the world's tropical rainforests, are afraid that the treaty might limit their ability to log and export timber or to clear land for agriculture.

Europe

According to many Europeans, part of the problem is that most nations speak of economic growth as the only alternative. But many in Europe feel that the human enterprise might just have grown as big as it can get. The governments of Europe are vastly influenced by their green parties and are determined to push environmental issues. European populations are now stable or shrinking. Constant growth is not the imperative it once was in the Old World.

The Skeptical Environmentalist

Not all Europeans are environmentalists. Bjorn Lomborg, director of the Environmental Assessment Institute in Denmark and author of "The Skeptical Environmentalist," feels that environmentalists exaggerate the negative state of the environment. They claim that natural resources are running out; population is growing, leaving less to eat; forests are disappearing; the planet's air and water are getting even more polluted.

Lomborg counters that natural resources are increasing, there is more food now than at any time in the world's history, and species are becoming extinct, but not as fast as environmentalists claim. He believes that most forms of pollution are associated with the early phases of industrialization.

He thinks that the problems of the world will best be solved not by restricting economic growth, but by accelerating it. He believes it's far more expensive to cut carbon dioxide emissions radically than to pay the costs of adapting to increased temperatures. He believes that for what it costs the United States alone to comply with Kyoto, the United States could furnish every person in the world with basic health, education, family planning, water, and sanitation services.

Lomborg claims focusing on development and industry in the third world rather than conservation will yield the best return in the long run. According to Lomborg, only when people are rich enough to feed themselves regularly do they worry about the effect their actions might have on their fellow man and future generations.

Lomborg has a supportive following on many talk radio shows, but few supporters in international scientific communities.

What about industry?

(Photo by Michael Tennesen)

100 Nobel Prize Winners

In May 2002, 100 Nobel Prize winners, including recipients from every year between 1972 and 2001, issued a statement on the 100th anniversary of the Nobel Prize in Oslo, Norway. They claimed that global warming originating from the wealthy few would most gravely affect the fragile ecologies of the poor and the disenfranchised, the majority of which live a marginal existence in equatorial climates. They believe that this injustice represents the most profound danger to world peace in the coming years. They stated "it is time to turn our backs on the unilateral search for security, in which we seek shelter behind walls. Instead, we must persist in the quest for united action to counter both global warming and a weaponized world."

U.S. Politics

When 25 nations gathered together in Montreal in 1987 to solve the ozone problem, global atmospheric science was enjoying a respite from politics in the United States. It's true that through the 1960s and 1970s, questions about pollution had pitted the rising environmental movement against some sectors of industry, but it seems that by the 1980s many of the old adversaries were on the same side. This was evident in the unilateral approach that was taken against ozone depletion.

Then conservative Ronald Reagan hailed the agreement as "a model of cooperation." But somewhere along the way environmental concerns have become an issue of liberal versus conservative, and this is the tragedy. How does an effort to better understand the global climate and the effects that we might be having on it become political? How does it evolve from a "we" issue to an "us versus them" one?

George Bush Sr.

Bush Sr. was a sportsman hunter who supported environmental issues, including the Clean Air Act. It was Bush Sr.'s credit-trading system that was used as a model for the credit-trading system in the Kyoto agreement. Bush Sr. was able to cut acid rain in half by allowing U.S. utilities to trade sulfur dioxide credits. At the 1992 Earth Summit, Bush Sr. made the commitment to cut emissions to 1990 levels by the year 2000.

Bill Clinton

Clinton rode the fence on global warming. He supported the environment, but also supported world trade and economic growth. He favored Kyoto, but under his administration, a time of economic prosperity, little was done to reduce greenhouse gas emissions. In fact, greenhouse gases increased significantly during the 1990s. Al Gore, his vice president, took a dominant role at Kyoto, but Congress never ratified the agreement.

George Bush Jr.

Bush Jr. saw too many flaws in the Kyoto agreement. Chief amongst them was the fact that developing countries like China and India didn't have to make any commitment. Some felt that by pulling out of Kyoto, Bush was simply stating the reality of U.S. politics toward the whole affair. Even though Clinton had backed it, nothing had happened. Why deny reality? The force of economics was against it.

Bush announced his own emissions reduction plan that tied greenhouse gas emissions to the *intensity* of the economy. To many environmentalists, that was actually an increase disguised as a reduction. Bush increased taxpayer subsidies for coal and proposed legislative rollbacks to the Clean Air Act, which his father had championed.

Despite a report from the National Academy of Sciences, agreeing with the basic IPCC assumptions, the Bush administration asked for another study. It proposed a decade of further research before anything beyond voluntary measures to reduce emissions from industry and autos could be considered.

In truth, there is reason to be skeptical. The range of climate warming projected by the IPCC over the twenty-first century is broad: 2.5–10.4°F (1.4–5.8°C). If the climate warms only 2.5°F (1.4°C), it might not be so bad. We could live with it. It might even be good for your roses. It might even help your vegetables to grow. But if the climate increases 10.4°F (5.8°C), that would spell major catastrophe: floods, droughts, hurricanes … the works.

The problem is that many want to forestall action, especially expensive action, based on early indications. But early indications are all we have. The earth has warmed 1.1°F (.6°C). The trouble is, if we wait for further climate change before we act, the changes we inherit will be ours for 100 years.

Many liken it to insurance. We take out insurance on our cars and our homes in the event of major catastrophes. We don't expect them. We don't anticipate a major fire in our home or a major accident in our cars, but we want to be prepared. Isn't it about time we took out some insurance against global warming?

The Cause

When I first heard about global warming, I thought, "Isn't this neat." I had written a number of articles in major magazines on the problems of preserving open space, saving our wildlife, reducing pollution, energy conservation, and tropical rainforest preservation. I felt then and feel today that these are all worthy goals in and of themselves. But global warming brought them all together and gave them global significance. An international agreement against global warming would put the whole world on the side of the environmentalists.

My first reaction was one of skepticism. This is too good to be true. I knew that many scientists, particularly those working in the environment, would be more likely to buy into it since it would serve their cause. Only after I started studying the evidence did I lose my skepticism. But even if we just forget about all the evidence concerning global warming, are pollution reduction, energy conservation, and rainforest production worthy goals to pursue?

Pollution Reduction

The burning of fossil fuels is the major contributor to increases in greenhouse gas emissions in the atmosphere. But it's also the major contributor to pollution. When we burn gas, oil, coal, and wood we release a bevy of pollutants into the atmosphere, many of which have greenhouse gas effects. Though carbon dioxide is not poisonous to our atmosphere, many of the other gases associated with it are. If we decrease the amount of fossil fuels we burn, we will decrease those pollutants and increase the purity of our air and water. The European Union estimates that between 43,000 and 180,000 cases of chronic bronchitis and between 15,000 and 34,000 cases of premature death could be avoided in the European Union by 2020 if all the nations in the EU adhere to EU environmental regulations and Kyoto standards.

Energy Conservation

We live in a world run on fossil fuels. We pay huge environmental costs for their extraction. The landscape is pockmarked with oil derricks, refineries, and vast open pit mines. Tankers like the Exxon Valdez off Alaska and the Prestige off Spain sometimes break up and leave environmental nightmares. And increasingly, oil is a factor in international warfare. We are fighting a battle over a diminishing resource. If we now begin an aggressive campaign to develop alternative natural resources we might even have a little oil left over. The technologies are there; all we need is the proper incentives to get going. The best incentive would be a public that is tired of the consequences of our reliance on fossil fuels and anxious to move on to the next phase.

Rainforest Preservation

Though about three quarters of the increase in greenhouse gas emissions are due to fossil fuels, a quarter is due to deforestation and the clearing and tilling of the landscape. Europe has already cut down most of its natural forests. North America has destroyed much of its own. The biggest virgin forests in the world range over the tropics in South America, western Africa, and southeast Asia. According to Wildlife Conservation International, these are hot spots of yet-to-be-discovered wildlife species.

But these forests are in danger. About 80 to 90 acres of tropical rainforest are destroyed every minute. Preserving these forests is a worthy cause. But unless the wealthy nations of the world are willing to help the third world, the forest will disappear and so will the enormous wealth of wildlife they contain.

The Least You Need to Know

♦ Voluntary measures to control greenhouse gas emissions have thus far proved ineffective.

♦ There are legitimate pros and cons to the efforts to do something about greenhouse gas emissions.

♦ U.S. politics have gone back and forth over the issue despite global concerns.

♦ Pollution reduction, energy conservation, and rainforest preservation are worthy goals themselves.

♦ Despite the skepticism, it might be time to take out some insurance against global warming.

Chapter 20

Summing Up the Argument

In This Chapter

- The historical evidence
- The role of carbon dioxide
- The havoc global warming is playing now
- What global warming will do in the future

Global warming and climate change are complex topics that bring together a number of different sciences. The issues are vast and complex. The fix is costly. The evidence is murky, and there is a lot of speculation. In this chapter, I will try to restate the case succinctly, review the most important facts, state the most current evidence, and discuss the scientists, their science, and their predictions.

Unraveling the Climate

I'm standing in the Laboratory for Tree-Ring Research under the football stadium at the University of Arizona in Tucson. Thomas Swetnam, the bearded director of the center that studies tree rings for clues to past climates, holds a piece of wood from a dead giant sequoia tree in front of me that is about three feet long, four inches wide, and a half inch thick.

It's a section from the giant sequoia that was cut from the bark to the center of the tree. It displays the annual rings of the tree as half moon lines spaced tightly together. On this piece of wood are pins with little quarter-inch flags that have dates on them. The flags start at 1958 C.E. and go all the way back to 523 C.E. Swetnam points to a section of the rings in the center of the board that are labeled as the late 1300s. Here the rings are closer together than elsewhere on the board.

Global Warnings

Scientists at the Laboratory for Tree-Ring Research are able to tell a lot about both temperature and precipitation by the width and the density of the tree rings. Trees generally put on fatter rings when it's wetter or warmer than normal. Reading rings of a tree at the highest place that they grow on a mountain will tell you how temperature varied. Reading the rings from a tree at the lowest place they grow will tell you precipitation levels.

"That's what got the Anasazis," says Swetnam. He explains to me that from about 800 to about 1200 C.E, the Anasazi Indians of the Southwest desert built elaborate cities in a number of places in Arizona, New Mexico, and Colorado. "But then in the late 1300s they ran into what the tree rings tell us was a long drought. They had to abandon those fabled cites and move south along the Rio Grande River."

The Laboratory for Tree-Ring Research was originally developed back in the 1920s to date these great Indian ruins, and they found they could create a calendar by matching the tree rings in the wooden beams of the ruins. Now the laboratory has joined with other scientists who are looking at a number of past climate signals including tree rings, ice cores, and marine sediments. They're trying to unravel another human mystery, as to whether we are headed for a similar disruption in our society as the Anasazis encountered, only this time due to global warming.

The Common Belief

Most scientists now believe that man is changing the climate of the planet by his emissions of greenhouse gases. The controversy is over the amount we're changing it and over the high costs of repairing the damage. Recently, scientists at the Laboratory of Tree-Ring Research put together a compilation of global temperature estimates of the last 1,000 years. The estimates were derived from a number of sources, including thermometers, tree rings, ice cores, and marine sediments. The resultant graph from all these disciplines looks like a hockey stick, where the end of the graph depicts a sharp rise, which represents the increase in temperatures over the last 150 years.

This chart is one of the lead graphs in the 2001 report from the IPCC on global warming. More than 2,000 scientists put together that report based on more than a decade of research. Though not every scientist signed off on every word, they did reach a consensus. They estimate that if the current rise of carbon dioxide in the atmosphere continues, the world will warm up from 2.5–10.4°F (1.4–5.8°C) by the end of the twenty-first century. The resultant hurricanes, floods, droughts, and sea-level rise could drown much of Florida, trash our agriculture, and cause extensive damage and disease.

More floods.

(Photo by Michael Tennesen)

Historic Gases

It was French mathematician Jean Fourier who in 1827 discovered that certain gases in the atmosphere act like a greenhouse. They let in sunlight—like a greenhouse does—but prevent some of the sun's warmth from radiating back into space.

The greenhouse effect is the consequence that certain gases have on the earth's climate, chief amongst them carbon dioxide. Though they constitute less than 1 percent of the atmosphere, they act like a blanket covering the earth. Without the greenhouse effect, the average global temperature would be around 0°F instead of the current toasty 59°F.

It wasn't until the 1930s that anybody noticed that there were problems with what scientists had come to call the greenhouse effect. Then, during a period of particularly hot summers in Europe, George Callendar, a British coal engineer, discovered that the world was gradually getting warmer. He pointed to rising levels of carbon dioxide as the culprit.

Charles Keeling was the first to develop an instrument for measuring CO_2 in the atmosphere. In 1958, during the International Geophysical Year, he took it up to the top of Mauna Loa, a high volcano on the island of Hawaii. There he recorded the rise and fall of CO_2 in the atmosphere throughout the seasons as plants grew and decayed. He left his instrument up there, put a few more out in other parts of the world, and discovered that CO_2 was gradually increasing in the atmosphere.

Global Warnings

Charles Keeling, a California scientist, was the first to document the rise of carbon dioxide in our atmosphere. Prior to the Industrial Revolution, which began in England in about 1750, the concentration of carbon dioxide in the atmosphere stood at about 280 parts per million (ppm). In 1958, when Keeling first measured it, it was 315 ppm. By 2001 it was 370 ppm. Scientists expect that the concentration of carbon dioxide in the atmosphere will double over pre-industrial levels to 560 ppm sometime during the twenty-first century.

Murky Thermometers

Scientists then started looking at the temperature, to see if there was a correlation. It wasn't easy. Historical temperature records stretch back 300 years in a few places, but only a few decades in most. Even in places where we've been keeping records, if you put your thermometer in the middle of a town and a city grows up around it, the asphalt and concrete buildings make it warmer. It's called the "urban heat island effect."

Warm Words

Proxies are things like tree rings, ice cores, and marine sediments, which scientists can analyze for indirect readings of historical temperatures of the atmosphere.

Satellites have made the readings of the earth's temperatures more accurate, but historical records are lacking. To get an idea of what temperatures were like in the past, scientists have had to look at certain *proxies*, like tree rings, ice cores, and marine sediments, which they can analyze and extract readings of historical temperatures.

Calibrating the Computers

One of the main reasons for studying the past is to help scientists predict the future of the atmosphere. Scientists use basic computer models for determining the weather and turn them toward the distant future. To do this, the programs are simplified, so that instead of telling us weather patterns in specific places, they tell us general climate parameters such as temperature. This is necessary because weather programs test the limits of today's supercomputers just to predict the weather 10 days ahead. To give us a picture of the climate 100 years out, you can only ask the computer simple questions.

Scientists have had some luck predicting El Niños almost a year in advance, but predicting climate 100 years out is untested science. To verify their accuracy of the models, scientists run their computer programs on past climate parameters and see if they predict the known outcomes. Our studies of past climates help us calibrate the computers that we then point toward the future.

Tree Rings and the Climate

At Tucson they've built a chronology of tree rings from pencil-width cores taken from the bristle cone pines in the White Mountains and the giant sequoias in the Sierra Nevada Mountains. These trees are some of the oldest on earth. One calendar that the lab put together of bristle cone pines went back 9,000 years.

During those years there have been a number of climate events that have affected man. A warming trend from about 950 to 1250 C.E. may have been a welcome respite in the United States, but it also made it possible for the Vikings in the Old World to rape and pillage Europe. The Vikings discovered America at that time and established a settlement in Greenland. But when a cooling trend known as the Little Ice Age came along about 1400 C.E., Viking settlers in Greenland brought their farm animals into their homes. When winter got too cold they ate the farm animals. They eventually died out when no boats could land on their frozen shores.

The changes in the climate over the last 2000 years—though they were rough on the Vikings—have represented slow one-degree shifts in the global temperatures. For the last 10,000 years the climate has been fairly stable. Prior to that it was a different picture.

Ice Cores and Ocean Currents

To look back further into the history of the atmosphere, scientists have gone to both Greenland and Antarctica, where they've uncovered a record of both the temperature and the atmosphere buried in the ice. Though a portion of the world's permanent ice is locked into glaciers in high mountain regions around the world, fully 99 percent of the earth's ice is in the great ice sheets that cover most of the island of Greenland and nearly the entire continent of Antarctica.

From 1989 to 1993, U.S. and European scientists spent five summers drilling a 2-mile (3.2 kilometers) core into the ice at the center of Greenland to get readings of the atmosphere over the last 110,000 years.

Climatoids

Snow in Greenland doesn't melt, but accumulates on the surface and compacts under the weight of new snow accumulating above. Each annual layer is about 12 inches deep toward the top of the ice sheet and about 4 inches deep at the bottom of the 2-mile (3.2 kilometers) ice core. The low-density, coarse-grained layers formed under the summer sun show up as light bands, while the fine-grained snow packed by winter storms appears darker. Scientists can read these cores on a light table.

Inside the ice cores are samples of the ancient atmosphere. The air in the ice congregates into bubbles and is further compacted, but upon release is a fairly accurate representation of the atmosphere at the time the ice was formed.

Scientists determine the temperature at the time the snow fell by measuring the amount of heavy oxygen (oxygen with an extra neutron or two) in the compacted snow. It's the first thing to fall out of the clouds when it gets cold and the amount of it in the ice is directly proportional to the temperature. The bubbles can tell you what was in the atmosphere when it was formed.

The winds of the world do a good job of mixing the atmosphere. If you were to release a large quantity of gas in Seattle, it would only take a few years before people in Bangkok could measure what you've released. Dust and sea salt don't mix well, but most gases are in the atmosphere long enough to mix globally.

The Greenland ice cores give us a marvelous record of the atmosphere that stretches back 110,000 years. What they show is that for the last 10,000 years the global climate has been very stable. The globally averaged temperature has remained about 58°F (14.4°C), except for the last 150 years when it rose to 59°F (15°C). The Little

Ice Age that froze the Vikings out of Greenland was only about 1°F (.5°C) cooler. The present amount of global warming has made us about 1°F (.6°C) warmer.

We are currently in an interglacial, a warm period between ice ages. The cycle has occurred every 100,000 years for some time now. Over the 110,000 year record of the Greenland ice cores there have only been two stable periods, the one we're in now and the one that occurred in the middle of the last ice age when central Greenland was about 40°F colder than today.

The rest of the time, the earth's climate was a roller coaster, swinging wildly back and forth. You step out your door one year and it's southern California. A decade later it's Vancouver. Another 10 years and it's Acapulco. From the end of the last Ice Age until the present, it's been really stable, except for a couple of dips.

Younger Dryas

The best-studied dip was the Younger Dryas event. It occurred about 12,800 years ago. Back then, the world was warming up out of the last Ice Age rather nicely when all of a sudden it took a big dive. Europe was about 12–16°F (7–9°C) colder than today—and it stayed that way for 1,200 years. Scientists wonder if global warming could trigger another such event.

The Younger Dryas was hell on nature. During the event, the mastodons, mammoths, horses, and saber-toothed tigers all went extinct. Early man contributed to this extinction through aggressive hunting, but scientists feel the climate was a major factor as well. When Dryas ended, it did so in a series of jumps, with central Greenland warming up 15°F in a single decade or less.

Climatoids

The Younger Dryas event is named after the dryas plant, which is a little rose found only in the tundra. During the Younger Dryas, the tundra reached all the way south into Virginia.

Scientists believe that the Younger Dryas event may have been caused by a shut-off of the Thermohaline Circulation (THC), a conveyor belt–like current in the deep sea that connects the major oceans of the planet. In this current, warm salty Atlantic water near the equator moves north toward Greenland and Labrador, where it has cooled enough to sink. The warmth that this current brings is the reason that Europe is as warm as it is, though it is in the same latitudes as Canada.

The sinking occurs in a couple of relatively small places in the sea, diving more than a mile deep before the water flips around and heads south down the Atlantic. The high salt content is necessary if the waters in the current are going to make the dive into

Climatoids

The Younger Dryas might have been created when ice water dams which had held back the fresh waters of the Great Lakes during the last Ice Age suddenly broke and inundated the North Atlantic with an infusion of fresh water. This diluted the salinity of the water and shut off the global ocean conveyor belt known as the Thermohaline Circulation.

the North Atlantic and turn south. Scientists feel that global warming could create enough ice melt and rain to dilute the water in the current and shut off the circulation. If it did, Europe and other parts of the northern hemisphere could get quite cold while the southern hemisphere could get hotter.

Scientists believe that the climate might be like a flip switch as opposed to a dimmer. If you put pressure on the switch gradually, it doesn't immediately start to change; it hesitates until there is enough pressure and then flips all of a sudden into a completely different state. Once there, it might just remain for centuries. It's happened before.

The Rise and Fall of Carbon Dioxide

Carbon dioxide might be a critical part of that switch. The Greenland ice cores give us a picture that goes back 110,000 years, but cores drilled in Antarctica give us a picture that goes back over 400,000 years. What Antarctica has shown us is that during the last four glacial cycles, the atmospheric concentration of carbon dioxide has varied from 200 ppm during the Ice Ages to about 280 ppm during the present interglacial. Over the last 10,000 years it's remained steady at 280 ppm, until the Industrial Age, when it started going up. It's currently about 370 ppm.

What does that say? Well, the difference between the Ice Ages and the interglacials is only 80 ppm of carbon dioxide. Though scientists don't believe that carbon dioxide caused the Ice Ages, it amplified the effect.

Snowball Earth

About 600 million years ago, an ice age got out of hand and froze the entire planet during a period called Snowball Earth. It seems that once ice starts forming, it may perpetuate itself by reflecting sunlight. Scientists who've studied this period of the earth believe that the worldwide global temperature average got as low as $-58°F$ ($-49°C$). The entire ocean froze to a .5 mile (1 kilometer) thickness. Life, which was pretty microscopic back then, was nearly wiped out.

Hothouse Earth

But carbon dioxide kept accumulating in the atmosphere, put there by spewing volcanoes. It increased 1,000-fold over 10 million years, and Snowball Earth turned into Hothouse Earth. During Hothouse Earth, the average global temperature soared to more than 122°F (50°C).

What all this shows is how important the content of carbon dioxide in our atmosphere is to climate.

The Current Picture

So what do we have now? What we've got is a period during which man has affected the atmosphere on a more rapid scale than has ever been known in nature. According to the World Meteorological Organization, nine of the ten warmest years on record have occurred since 1990.

◆ **Disappearing ice**   Though scientists predict that the biggest stuff is yet to come, some changes are already evident. There has been a widespread retreat of mountain glaciers outside the poles just in the last century. Ice in the European Alps has declined by half. There's been a 40 percent decline in late summer Arctic sea-ice thickness.

◆ **Glacierless National Park**   When naturalists first hiked through Glacier National Park more than a century ago, there were about 150 glaciers nestled into its high cliffs and jagged peaks. Today there are only 35. In another 30 years, scientists estimate even those will be gone. Glacier National Park will be glacierless.

◆ **Warming up to Russia**   The United Nations Environmental Program recently completed a survey of glaciers in the Himalayan Mountains and found that dozens of lakes were so swollen from melting glaciers that they could burst in the next few years, inundating villages throughout the region. The collapse of the Maili glacier on the northern edge of the Caucasus Mountains ripped out trees and tossed massive trucks like toys into the air. It buried a Russian village under 3 million tons of ice and mud and left at least nine dead.

◆ **Antarctic ice**   Four Antarctic ice shelves have collapsed, including the Wordie Ice Shelf, which measured 772 square miles in the 1940s but has shrunk by two thirds today. The Prince Gustav ice shelf disappeared altogether. The most dramatic of these collapses was easily that of the Larsen Ice Shelf. As scientists watched via satellite in January 1995, it disintegrated over the span of five days.

At one point a slab of ice the size of Rhode Island shattered free and floated away. David Vaughn of the British Antarctic Survey commented that it was as if a godlike hammer had fallen on them. There have been other collapses as well.

Over the last 50 years, winter precipitation in the Sierra Nevada and Rocky Mountains has been falling more and more in the form of rain rather than snow. The Southwest desert depends on the snowpack to continue to deliver water late into summer. With more rain and less snow, the summers will be getting drier.

Environmental Disasters

Nature generally stands to lose the most from global warming. Plant life and wildlife don't have air conditioning. Generally, plant life evolves slowly with the climate. If the climate gets progressively warmer, then trees and plants will migrate out of our parks and move northward. The trouble is that roads and cities break up the natural migration paths. Plants and trees are more likely to die out along with the wildlife they support rather than migrate.

Polar Bear Jail

In Churchill, Canada, 1,500-pound polar bears which used to feed on seals out on the frozen Hudson Bay grow hungry as the ice on the bay melts earlier each year. The bears are showing up in town much more frequently. Though the Canadian Wildlife Service tries to trap them and helicopter them away, if civilian encounters with polar bears continue, it could lead to catastrophe.

Starving Whales

Only five years after being taken off the endangered species list, the gray whales which migrate from the Bering Sea to Baja California each year have been dying off in large numbers. The population has dropped from an estimated peak of 26,635 whales in 1998 to 17,414 whales in 2002, a decline of more than a third. The 35- to 50-ton beasts spend their summer in the Bering Sea gorging on millions of amphipods—crustaceans that live in tubes on the shallow ocean floor.

Though the whales migrate all the way from the lagoons off Baja California to the Bering Sea to eat, the amphipods rely on their food being brought to them by ocean currents. But global warming may be altering ice patterns and ocean currents in ways that prevent plankton from making its way through the water column to the amphipods. The whales, which feed on these shrimp-sized creatures, could be the largest victims yet of global warming.

IPCC Predictions

So what can we expect for the future? The Intergovernmental Panel on Climate Change (IPCC), a group of over 2,000 scientists appointed by the United Nations and the World Meteorological Organization, concluded in their 2001 report the climate will rise between 2.5–10.4°F by the end of the twenty-first century. The panel is made up of the superstars of climate research—university department heads and leading government agency researchers from all over the world. And they've been studying this stuff since 1988.

If the temperature only rises 2.5°F, we might be able to live with it. If it rises 10.4°F, then that's going to cause major problems. Ten degrees is as great a change as ended the last Ice Age. Only instead of happening over about 13,000 years, it will happen in less than ten decades. Some effects are expected to become apparent in the next 30 years.

Sea-Level Rise

The IPCC predicts that sea level will rise in the range of 4 inches to 3 feet (.09 to .88 meters) over the next century. The greatest impact will likely be in recently developed coastal areas like Florida and parts of the U.S. Gulf and Atlantic coasts. In Florida, sea-level rises of only a few inches will advance the shoreline as much as 400 feet in low-lying areas. This will flood shoreline homes and hotels and erode the state's treasured beaches. Sea-level rise may also damage the Everglades, offshore islands, and coral reefs. In short, much of what has made Florida so precious may be lost in a globally warmer world.

> **CAUTION**
>
> **Global Warnings**
>
> Global warming might increase forest pests in Florida and throughout North America. This could weaken the trees and make the forests ripe for fire.

Going to Death Valley

A warmer world is likely to be a wetter one with catastrophic storms and floods over much of the globe. On the other hand, some areas like the plains and the deserts will get hotter and drier. Parts of the plains will turn into deserts.

More deserts.

(Photo by Michael Tennesen)

In the Southwest desert, where temperatures reach between 110–120°F (43–48.5°C) in the summers, temperatures could start climbing up to 120–130°F (48.5–54°C). To get an idea of what that feels like, you have to go to Death Valley, California.

Death Valley is the hottest and driest place in the western hemisphere. On July 10, 1913, the mercury there reached 134°F, a world record for many years. When it gets above 115°F (46°C), the visitor's center at Death Valley National Park wraps its door handles in Styrofoam so people won't burn their hands, and the maintenance people keep their tools in water so they won't burn theirs. People use gloves to drive their cars so the steering wheels won't burn them. Locals shut their water heaters off in late spring since the tap water is plenty hot enough.

Late July is the normal peak of yearly temperatures, with temperature readings regularly approaching 130°F (54°C). Around that time there are lots of electrical outages, where the air conditioning goes off for as much as 4 or 5 days at a stretch. Then the park rangers sleep under wet sheets with wet hair. Or they sleep on picnic tables outside with the sprinklers on.

Park rangers describe Death Valley in July as being similar to roasting a turkey and you've just opened the oven to a blast of hot air. And the door is stuck and so are you.

The Costs

Fixing global warming is not going to be cheap. The Kyoto agreement, which would have required the United States to reduce its greenhouse emissions 7 percent below 1990 levels, was only supposed to be a first step. Entirely stopping climate change will require 50 to 60 percent cuts. This is likely to be hugely expensive, requiring major reductions in the use of gasoline, fuel oil, natural gas, and coal.

Doing something now about global warming is like taking out insurance. We take out insurance on our cars and our homes in the event of major catastrophes. We don't expect them, but we want to be prepared.

Maybe it's time to contact an agent.

The Least You Need to Know

- ◆ Greenhouse gases like carbon dioxide keep us from freezing to death.

- ◆ The trouble is, we've increased its concentration in the atmosphere by burning fossil fuels.

- ◆ Scientists have studied tree rings, ice cores, and marine sediments to get a picture of our climate in the past.

- ◆ Scientists use these studies to help them predict the future.

- ◆ Glacial ice is melting all over the world as a result of the current rate of global warming.

- ◆ If we don't do something quickly, it might start feeling like we're living in Death Valley.

Part 6

The Solution

Okay, now that I've tackled the basics, discussed the politics, and laid out the pros and cons, it's time to address the problem. This isn't going to hurt. It might save you some money. And it might save the world some misery.

I'll start by looking around the house and discussing all the ways we can save energy. Some are simple fixes that won't cost anything. Others might require some investment, but you'll be saving in the long run on your electric and heating bills.

Then I'll invite you to take a look at your car. You don't have to go out and sell your eight-cylinder sedan today, but think about your options for future purchases. Do you really need that eight-passenger SUV or could you and the earth live better and healthier on a few less horsepower?

Finally, I'll take a look at some energy alternatives. With gas reserves on the wane, it's prudent to investigate. There are ways to make our fossil fuels cleaner right now. Plus there are some exciting possibilities in solar, wind, and fuel cell technologies. It's time to make some changes.

Chapter 21

It Starts with the Individual

In This Chapter

- ◆ It's all about saving energy
- ◆ Conserving energy in the home and office
- ◆ Recycle and reuse
- ◆ How we get our electricity

Reversing global warming starts with the individual. It's a two-stage process. First you have to make a decision that it's a problem worth addressing. Then you have to make a commitment to do something about it. Most of the greenhouse gases in the atmosphere come from the burning of fossil fuels. You and I do this by using energy at home, in the office, and in our cars. I'll address the issue of cars in the following chapter. In this chapter, I'll talk about how we can cut back on energy usage where we live and where we work. I'll also talk about how we get our energy and what we can do to influence the cleaner generation of electricity.

In and Around the House

If you're sitting in your easy chair reading this book in a room with all the lights on, miffed about the draft coming through your closed window, thinking about checking that old refrigerator for something to eat, and looking forward to taking a nice, long, hot shower ... have I got some news for you. You might just be a walking, talking, global warming machine. But don't worry. I've got some great stuff here for you to consider to cut into that unnecessary load of energy you might be about to consume.

The Windows

You're feeling a draft coming from the window, but it's closed. Don't worry. You're not losing your marbles, you're just losing precious heat through those flimsy portals of yours. Windows are typically the weakest point between your living room and the outdoors. The cheapest way to stop those drafts from running up your heating bill is to reduce the heat loss through the cracks and crevices around the window casings. Caulk around the edges of those drafty frames. Do it both on the inside and outside of your windows. Re-putty loose frames. Put some weather-stripping around the windows that are opened most often and around your doors. You've just gone a long way toward stopping heat loss through your windows.

> **CAUTION**
>
> **Global Warnings**
>
> The amount of energy that leaks through windows in American homes each year is equal to the energy produced by the annual flow of oil through the Alaskan pipeline.

Now look at the glass. If it's single-pane, it's not doing much more than blocking the wind. Double-pane windows provide an extra layer as well as some dead space between the panes of glass that can increase insulation dramatically. Storm windows add that extra layer as well. Most hardware stores will sell you plastic barriers that you can use to achieve the same effect during the winter months. They'll cut heat loss through single-pane windows by 25 to 40 percent. Curtains and shades are better than nothing, but not as good as double-pane windows.

Insulation

Insulation can turn a drafty old house into a cozy home that is much cheaper to heat or cool. The first place you'll want to consider insulating is the ceiling or attic. It's where the majority of heat is lost or gained. The next most efficient place is the walls. After that it's beneath the flooring. Insulation comes in a variety of forms including loose-fill cellulose (made from recycled newspapers), fiberglass, foam board, insulated panels, and sprayed-on cellulose. The difference in the warmth of my home the day before I insulated the attic and the day after was dramatic.

The Lights

Lighting uses about a fourth of all the electricity consumed in the United States. That includes exterior and public lighting—in-home lighting isn't as large, but we can cut the in-home lighting bill by 90 percent, if we use the most efficient sources—including a little sun. Using incandescent light bulbs is the most wasteful kind of light you can use both at home and in the office. About 90 percent of the energy it consumes is given off as heat, and only 10 percent is converted to light. Fluorescent lighting typically yields three to four times as much light for the same amount of energy.

Fluorescent lights are better than they used to be. The newer bulbs are much closer to natural sunlight than their incandescent counterparts. And new ballasts don't hum like the older ones. Fluorescent lights cost more than incandescent, but fluorescents can last up to 12 times as long. When you add the longevity of the fluorescent bulb to the savings you'll get on your electric bill, they end up being about half as expensive as an incandescent bulb.

Climatoids _____

Compact fluorescent lamps screwed into ordinary sockets can provide incandescent light color, last up to 12 times longer, and use one fourth of the electricity. Use them to replace your most frequently used lights. Each incandescent bulb that you replace will prevent 1,000 to 2,000 pounds (453 to 906 kilograms) of carbon dioxide emissions over the life of the bulb.

The Refrigerator

Refrigerators and freezers consume about a sixth of all the energy used in a typical American home, but a lot of that can be saved. First of all, if you've got an old refrigerator, get rid of it. My family saved almost $25 a month on our electric bill when we traded in our old refrigerator for a new one. That was enough of a saving to cover the payment on the new one. Advances in technology have made the modern-day refrigerator about 60 percent more efficient than it was 20 years ago.

In general, the simpler refrigerators are the most energy-efficient. Automatic ice-makers typically increase energy use by 14 to 20 percent. Water dispensers waste energy as well. Check the "Energy Guide" labels that appear on the new models and compare the one you want with the others. Getting the one that uses

Global Warnings _____

The costliest refrigerator you have is that extra one in the garage that's keeping that six pack cold. Pull the plug.

the least energy doesn't mean your food is going to spoil. It means your bills will be lower and the atmosphere cleaner.

Mark packages and containers in your refrigerator so you can find what you are looking for quickly. Standing there, scratching your head with the refrigerator door open, is a waste of energy.

The Heater

Think about getting a programmable thermostat that lets you set the heat for different parts of the house, and reduce it when you're not there or fast asleep. Setting the thermostat back 10°F (5.5°C) at night can save 10 percent on your heating bill. You can get yourself a real nice blanket with those kinds of savings. Or buy yourself a sweater and keep the heat lower during the day. Just turning the thermostat down one degree cooler than normal will reduce carbon dioxide emissions from electrical generating facilities by 1157 pounds (525 kilograms) a year.

Climatoids

Deciduous trees—the ones with the leaves—planted to the south, east, and west of your home will provide cooling shade in the summer. In winter when the leaves fall, they'll allow the sunlight in to warm your home.

Air Conditioning

The same goes for the air conditioning. Set your thermostat a little higher and you'll save energy. Don't turn on the air conditioning because you're too lazy to take off your sweatshirt. Try a ceiling fan instead of air conditioning. It's possible to cool your home without using a lot of energy. Install awnings on south facing windows where there is no roof overhang. Hang white window shades or blinds, which can reflect as much as 40 to 50 percent of the midday heat.

The Water Heater

Heating your water is the second-biggest use of home energy after heating and cooling. Heating water costs the typical family about $160 to $390 a year. If you use electricity to heat your water, then you are paying about twice as much as if you used natural gas or propane. Whether it's gas or electric, consider wrapping your water heater in an insulation blanket made especially for water heaters and you'll lose a lot less heat to the air.

Another way to lower your water-heating bill is to use less hot water. Energy-efficient showerheads can cut back on your water usage by half. They work by aerating the water or breaking it up into many tiny streams. You'll still get a good shower.

You can use aerators on your kitchen and bathroom faucets, save energy, and still get the job done. Retrofitting one showerhead and two faucets will reduce your annual carbon dioxide emissions by 580 to 3,200 pounds (261 to 1468 kilograms).

Climatoids

Turn your hot water down to 120°F (48.5°C) and your hot water costs will drop as much as 50 percent.

The Dishwasher

The dishwasher uses almost as much energy as a clothes dryer or freezer. Most modern machines have a booster to get the water temperature up to 140°F (60°C) to cut grease and kill germs. This isn't that bad, since it allows you to turn your hot water heater down to 120°F (48.5°C). The most efficient water heaters on the market currently use about 5 gallons of hot water per load and this is plenty. (Sailors aboard combat ships get two gallons to clean their whole bodies.) Consider using the *light wash* or *energy saving cycles* when you can. Use the least amount of energy to get the job done, not the most. Choose air drying over heat drying; it'll make a difference. And scrape your dishes off, don't rinse them.

If you are going to rinse your dishes off, why not just wash them by hand? We do around our house. We had a dishwasher, but we got rid of it. We found that it didn't take any longer to wash the dishes than it took to load the dishwasher. My mother, however, disagrees. But my electric bill is lower than hers.

The Stove

Cooking appliances are some of the biggest energy consumers in the home. If you've got the air conditioning on, then the hot stove will make it work harder. Gas is your best bet. The fuel is used directly to cook the food. With electric, the fuel is brought to a power plant, converted to electricity, and then shipped long distances over power lines. It requires three or four units of fuel to produce one unit of electricity.

Consider a microwave for certain cooking tasks. They use up to two-thirds less electricity than conventional electric ovens. Toaster ovens, crock-pots, and pressure cookers also save substantial energy costs in comparison with conventional ovens.

Electronics

Whether you are at home or in the office, keep your electronics off when not in use. Leaving your computer on doesn't make it run better, it only wastes energy. It also reduces the life of your screen and hard drive. The same goes for your TV and stereo.

If you want to leave a light on when you're not there to scare away prowlers, put it on a timer. It will look more natural (that is, it will scare more prowlers) and save energy as well.

When buying new electronics, look for the Energy Star label that will tell you the energy-efficient ratings. Energy Star was introduced by the U.S. Environmental Protection Agency in 1992 as a voluntary labeling program designed to identify and promote energy-efficient products, which would reduce carbon dioxide emissions. Energy Star has expanded to include new homes, residential heating and cooling equipment, major appliances, office equipment, lighting, and consumer electronics.

You Are What You Eat

You can also save energy and reduce greenhouse gas emissions by thinking about the food that you eat, too. Those grapes from Chile may taste great in January, but think about the fuel that was wasted getting them to your door. Buy local where you can. Farmers markets are great places to pick up fresh fruits and vegetables in season. Enjoy your meals in comfort knowing that a minimum of fossil fuels was used to get them to your door.

In the Garden

Not only do house and garden plants give off oxygen and absorb airborne toxins, they're good for our emotional health as well. Diane Relf, Ph.D., founder of the People Plant Council, a consortium of university horticultural therapists, says that just looking at plants and gardens can reduce blood pressure. In one study, patients with a view of nature got out of the hospital faster, needed fewer painkillers, and even bothered their nurses less. Here are some hints to make your garden more environmentally kosher.

Pesticides

Pesticides for the most part are petrochemicals—petroleum products. They utilize fossil fuels in all stages of their production. They're also poison, and birds, butterflies, pets, and small children are sensitive to them. And yet the National Audubon Society reports that homeowners use three times more pesticides per acre than the average farmer.

In general, try the least-toxic product first. If that doesn't work, try something a little stronger. But utilize the least amount of toxins to get the job done. For ants, fill a sprayer with a few tablespoons of soap, add water, and spray the ants. It kills them and cleans the chemical trails they leave to attract other ants. For mice or rats use mechanical traps rather than poisons. Bait the traps with peanut butter, which is more attractive to them than cheese.

Rat traps instead of poison.

(Photo by Michael Tennesen)

Spray your plants with insecticidal soaps, which kill insects but are not harmful to you or other animals. Soaps for both insects and funguses are made from plant materials and are available at most nurseries. Or simply wash bugs off with a stream of water. Do it early in the day, so the moisture will dry and funguses won't flourish. Flowering plants like daisies will attract predatory bugs and help contain pests. Or buy your own predatory bugs, like ladybugs, and release them in the gardens. They're available at many nurseries. Keeping your garden in a natural balance will produce much healthier plants and flowers.

Climatoids

To take care of snails, twist pie tins into the ground so the lids are even with the dirt, then fill them with beer. The yeast in the beer will attract the snails and they'll drown. Or come out an hour after dark with a flashlight, pick the snails up, and put them into a jar of soapy water. Ten minutes three nights in a row will kill the active snail population in most gardens.

Use native plants where you can. They need less water and are more resistant to pests and disease so you'll need less pesticide. Look for nurseries that specialize in them. Visit your local arboretum and scout up some good native possibilities. It's a little more trouble to find them, but a lot easier taking care of them.

Ladybugs instead of pesticides.

(Photo by Michael Tennesen)

Compost

Consider starting a compost pile in the corner of your yard. Instead of putting the garbage down the drain, put it on the compost pile and you will be helping keep our rivers cleaner. Pile grass clippings, weeds, and leaves on it and the garbage man won't have to burn energy to haul those items to the dump. And your plants will love the dark loamy mulch that comes out of it. There are classes available in many communities to show you how to get started. Don't put meat, fish, or any animal products on your compost pile, and the pile won't attract insects.

Putting a little manure or organic fertilizer on your compost will get it going and decaying. Water it once in a while. Also, take a shovel or spade and turn it over every month or so and it will break down faster. I've got a compost pile in my yard and I use the mulch I get for my roses. They are the best in the neighborhood.

Recycling

Reuse what you have and recycle things you can't use. By reusing things we save energy and resources needed for the future. We also reduce the greenhouse gas emissions and pollution that occur from producing something new.

Producing new paper, glass, and metal products from recyclable materials saves 70 to 90 percent of the energy, pollution, and greenhouse emissions that would result from using virgin materials. You can save a good-sized tree by recycling a stack of newspapers only 4 inches (10 centimeters) high. Support recycling efforts in your community and tell your local officials that you think it is important. Why throw away perfectly good resources in our landfills? We don't have enough space left. The typical American discards 1,300 pounds (590 kilograms) of trash each year, and we're running out of places to bury it.

CAUTION **Global Warnings**

> Every year more than 200 million trees are cut down to make U.S. newspapers, the majority of which are tossed into the trash. Americans discard enough aluminum cans annually to rebuild the entire U.S. commercial airline fleet four times over. Buy products with reusable or recyclable packaging instead of those packaged in non-recyclable materials. You'll save natural resources and reduce your greenhouse emissions by 230 pounds (104 kilograms) a year.

The Price of Coal-Fired Energy

Electrical generation from fossil fuels, particularly coal, is the single largest source of greenhouse gas emissions and air pollution in the United States. Coal-fired power plants emit more than 31 percent of the nation's carbon dioxide. But as I mentioned in Chapter 19, the burning of fossil fuels and the production of carbon dioxide comes with a generous portion of fellow pollutants as well. Coal-fired electrical generating facilities account for more than 60 percent of the nation's sulfur dioxide emissions, 23 percent of its nitrogen oxide emissions, and 32 percent of its mercury emissions. Let's take a look at some of the more deadly problems that coal-fired power plants emit.

Climatoids

In the United States today, coal is the number one source of electricity produced (54 percent), followed by nuclear power (21 percent), followed by hydropower (16 percent), natural gas (9 percent), oil (2 percent), and other non-renewable sources (3 percent).

Acid Rain

Acid rain is formed when sulfur dioxide and nitrogen oxides react with water and oxygen in the atmosphere to form acidic compounds. These compounds fall to Earth in the form of acid rain, snow, sleet, hail, and even fog. It's measured by the pH scale, which runs from 1 (highly acidic) to 15 (highly alkaline), with 7 being neutral. Rain has a normal, slightly acidic pH of around 5.6, whereas acid rain is usually between 4 and 5 but can be as low as 1.5. The deep South and the Midwest are home to the largest number of coal-fired power plants, but they may escape the worst acid rain damage. That's because damage drifts eastward and falls on the eastern coasts of the United States and Canada.

Acid rain ruins lakes. Insect populations that surround acidic lakes are much less diverse than those around normal lakes. One study found that a number of important groups of insects—including dragon flies, damselflies, caddis flies, and mayflies—disappear when the pH of a lake falls below 5.5.

This affects fish populations. Scientists have identified 200 lakes in the Adirondacks and 200 in Ontario where all fish—including trout, walleye, and bass—have disappeared because acid rain made the waters too acidic for them to survive. The same goes for frogs, toads, salamanders, and other amphibians. When insects, fish, and amphibians disappear, so do the animals that rely on them for food. The acidification of our natural waters hinders the reproduction of a number of waterfowl, including ring-necked ducks, mergansers, and loons. It also affects river otters.

Acid rain eats away our buildings and monuments, too. Limestone and marble, the stones that form many of the buildings and monuments in Washington, D.C., are especially vulnerable to acid precipitation because these monuments were built with materials that contain calcium carbonate, which dissolves easily in acid. Many of the exposed areas on these structures show roughened surfaces and loss of detail in the carvings. The U.S. Park Service reports that the buildings affected include the U.S. Capitol, the Smithsonian Institution, the Arlington Memorial Bridge, the Lincoln and Jefferson Memorials, and the Washington Monument.

Mercury

Coal-fired power plants are the largest source of mercury emissions in the United States, accounting for an estimated 32 percent of emissions. Mercury in the air eventually ends up in lakes, rivers, and coastal waters, where it is a deadly poison. Just one drop of mercury can contaminate a 25-acre lake, making fish unsafe to eat. Mercury can cause brain, lung, and kidney damage, reproductive problems, and even death.

> **CAUTION**
>
> **Global Warnings**
>
> Salmon, lake trout, and walleye have some of the highest amounts of mercury in Great Lakes fish. Since these are species that people like to catch and eat, they In low doses mercury can cause subtle but permanent brain damage to humans. In high doses it can kill. Mercury is especially dangerous for children and pregnant women. The young and the unborn are particularly susceptible.

Much of the mercury from Midwestern power plants ends up in the Great Lakes. The mercury is stored in the fat of animals and accumulates as it works its way up the food chain. In other words, insects that eat plants get one dose, the fish that eat the insects get a higher dose, and the fish that eat the fish get the highest dose of all. Predator fish such as salmon, lake trout, and walleye may have concentrations of mercury more than a million times higher than the surrounding waters.

Ozone

The natural ozone layer in the stratosphere shields the earth from harmful ultraviolet rays, but ground-level ozone is a powerful respiratory irritant that is particularly dangerous for the elderly, children, and people with asthma or other respiratory diseases. Ozone triggers more than 6 million asthma attacks each year.

Ozone destroys both trees and crops and is responsible for the haze at many of our national parks. Ozone interferes with the production and storage of starches in all types of plants. This reduces growth rates and weakens plants. They then become more susceptible to insects, disease, and other environmental stresses.

> **CAUTION**
>
> **Global Warnings** _____
>
> Researchers at Carnegie Mellon University in Pittsburgh studied the health effects of pollution from fossil fuels on death rates in four international cities including Sao Paulo, Brazil; Mexico City; Santiago, Chile; and New York. They found that the burning of coal and gasoline is causing people to die prematurely from asthma, heart disease, and lung disorders. They estimate that more people are killed by air pollution than by auto accidents.

Old Power Plants Kill

The passage of the Clean Air Act in the United States has cleaned up many power plants and reduced sulfur dioxide emissions. However, there is a loophole in the Clean Air Act that allows older, dirtier power plants to emit more pollutants than newer, more efficient plants. There are a number of plants in the South and Midwest that have used this loophole to make significant upgrades to their facilities without addressing the environmental issues. The Sierra Club estimates there are as many as 600 existing power plants between 30 to 50 years old that are 10 times as dirty as new power plants built today. Congress set up this loophole because they assumed that the older plants would be replaced by newer facilities, but power companies are using the loophole to extend the life of plants way beyond their years. It needs to be stopped. Let your representative know how you feel.

Alternatives to Coal

English lawmakers switched their country's electric power plants from coal to natural gas and as a result had no trouble meeting Kyoto standards. The same could happen in North America. There are new efficient technologies for natural gas power plants that require less energy to run and would save money over the lifetime of the facility. Natural gas is inexpensive and has the lowest carbon content per unit of energy of all the fossil fuels.

Abundant, clean, and reliable energy sources such as wind, biomass, and solar power are catching on in North America. Sacramento, California, announced a $26 million program to use solar power to run more than 3,000 homes, the largest solar project in the nation. Mid-American Utilities has opened the world's largest wind power plant in Buena Vista County, Iowa. The government needs to encourage more efforts like these through tax cuts, rebates, and other incentives to reduce our reliance on fossil fuels. Call, write, or e-mail your legislators and ask them what they are doing about this. If everyone who read this book did just one of these things just *once*, it would make an enormous difference.

The Least You Need to Know

- The production of electricity is one of the major sources of global warming in North America.

- You can cut down on greenhouse gas emissions by being more energy-efficient at home.

- Newer appliances are much more energy-efficient than older ones. Look for Energy Star labels on new products to determine which are the most efficient.

- You can reduce carbon dioxide emissions by recycling around the home and by composting in the garden.

- Electrical generation from fossil fuels, particularly coal, is the single largest source of greenhouse gas emissions and air pollution in the United States.

- Switching to natural gas and alternative energy solutions to generate our electricity could do a lot to clean up the air.

22

What About My Car?

In This Chapter

- ◆ Raising fuel economy standards
- ◆ The hidden costs of SUVs
- ◆ Hybrid and electric cars
- ◆ Tips on saving gas

Let's look at our motor vehicles—not the fronts, but the backs where the tail pipes are. Automobile and truck emissions are one of the largest sources of carbon dioxide in America today. In fact, every gallon of gas burned creates 28 lbs (12.2 kgs) of carbon dioxide in the atmosphere—19 pounds from the tailpipe and 9 from refining and transporting the fuel. The average car creates about 70 tons of carbon dioxide over its lifetime, the average SUV around 100 tons. Raising fuel efficiency standards is the largest single step we can take to curb global warming. It's time to get going.

What Would Jesus Drive?

When Congress failed to increase fuel efficiency standards in 2003, a broad coalition of religious leaders got together and formed the National Religious Partnership for the Environment. The organization immediately began to lobby Detroit auto executives asking for improvements in fuel economy. They claimed they had a biblical mandate to be good stewards of God's creations. They also said they felt a responsibility to the poor who were especially harmed by pollution.

Climatoids

Current U.S. standards require new cars to average 27.5 miles a gallon (mpg), while light trucks, sport utility vehicles (SUVs), and vans must average 20.7 mpg. By contrast, the current standard in Japan is 30.3 mpg and in Europe 33 mpg. There are no breaks for SUVs.

One of the smaller groups within the religious partnership, the Evangelical Environmental Network, decided they would take it a step further. They purchased TV spots and ads in national magazines that posed the question "What Would Jesus Drive?" For that matter, what would any of the great prophets—Moses, Mohammed, Confucius, the Buddha—drive if they were running around in the flesh today? Would they pull up in a Humvee, spouting fumes and sucking up gas? Probably not. So why should we?

Top Five Global Polluters

America's cars and light trucks alone produce 20 percent of the U.S. carbon dioxide pollution. The pollution is not only a threat to our atmosphere but a threat to our health. According to the American Lung Association, nearly 117 million Americans live in areas where the air is unhealthy to breathe. Light trucks and SUVs magnify the health risk by spewing up to three times more smog-forming pollution than regular cars.

U.S autos alone emit more carbon dioxide than the total output (autos, industry, agriculture—everything) of all but four countries. What this means is that if we were to rate the world's top five emitters of carbon dioxide, the list would look like this:

1. United States

2. China

3. Russia

4. Japan

5. U.S. autos

The more gas you guzzle, the more carbon dioxide you are going to emit. When you compare the total tonnage of carbon dioxide produced over an average car's 125,000-mile lifetime, you get a clearer picture of the problem. A Ford Excursion, which gets 13 miles per gallon, emits 134 tons of carbon dioxide. A Jeep Cherokee, which gets 18 mpg, emits 96 tons. A Honda Civic at 36 mpg emits 48 tons. A Honda Insight at

65 mpg emits 27 tons. The Ford Excursion emits 5 times as much carbon dioxide in its lifetime as does the Honda Insight.

The SUV and Light Truck Loophole

Back in 1975 when fuel efficiency minimums were first adopted under the Corporate Average Fuel Economy (CAFE) standards, the term "light truck" referred to a vehicle that was used to haul hay on the farm and lumber at the construction site. Lawmakers, reacting to the petroleum crisis in 1973 and growing amounts of auto exhaust pollution passed this legislation, which set the makers of American sedans on a course toward lighter and more fuel-efficient cars. The CAFE standards, however, included exemptions for light trucks, which were then used by farmers and workers that needed more powerful cars for their jobs.

But what happened is that American automakers used the light truck definition to skirt the law. Back when the law was first passed, light trucks, SUVs, and minivans made up only 20 percent of the vehicles sold in America. Today they make up more than half the vehicles sold. But these cars are more likely to haul lattes home from Starbucks than lumber from the yard. Though many are used as family cars, they are more often seen with single drivers than with multiple riders.

The hidden cost of SUVs is the price we pay with our natural resources. To keep the gas tanks of these cars full, oil companies seek to drill in new areas, including some of our nation's most sensitive wilderness habitats. These include the fragile coastlines of Florida and California, the lands around Yellowstone National Park, and the Arctic National Wildlife Refuge in Alaska.

The price of SUVs.

(Photo by Michael Tennesen)

The Exxon Valdez disaster off Alaska and the breakup of the oil tanker Prestige off the coast of Spain serve to remind us of the dangers of transporting petroleum products in and around our national and international treasures. Small spills and oil leaks, which never make the paper but are no less disastrous, happen all the time.

Global Warnings

Switching from driving an average car to a 13-mpg SUV for one year would, according to the Sierra Club, waste more energy than if you …

♦ Left your refrigerator open for 6 years.

♦ Left your bathroom light burning for 30 years.

♦ Left your color television on for 28 years.

Auto-industry advertising portrays SUVs as the ticket to the great outdoors. They are shown splashing through creeks, churning up mud, running along the beach, and climbing high, rocky mountains. In reality, most of them never see dirt unless they're tearing up the lawn next to the driveway.

SUV Safety

Though many people buy SUVs for safety purposes, the opposite is actually true. The height that makes the SUV driver feel above the traffic and safer is the Achilles' heel of the vehicle. When SUVs swerve to avoid an accident, they are four times more likely to roll over. Rollovers account for 22 percent of deaths in normal cars but 62 percent in SUVs.

Hot Debates

According to a government study in the 1990s, at least 2,000 traffic fatalities would not have occurred had their cars collided with another car rather than a light truck or SUV.

Because an SUV is taller, heavier, and more rigid, it's more than twice as likely to kill the other driver in an accident. That's because when an SUV and a car collide, the height difference, stiff frame, and greater mass of the SUV create a weapon that can be lethal to passengers. They're also lethal to pedestrians, bicyclists, and motorcycles. Braking standards for SUVs are weaker than for cars.

Measuring Savings

The truth is that automakers could make SUVs that were more fuel-efficient. It's just that the pressure is not there. People still talk about how *powerful* their cars are, how big an engine they've got, how many cylinders they have. You'll know that a big part of the battle against global warming will be won when you start hearing more people brag about how many miles per gallon (kilometers per liter) their car gets than about how fast or big it is. To help get this trend started, let's look at the Sierra Club's list of some of the most and least fuel-efficient vehicles in the different car categories for 2003. In general, manual transmissions are more fuel-efficient than automatic transmissions, so most of the cars listed here are manual (for automatic transmissions subtract 2 to 8 more mpg):

The most efficient cars

- Most efficient two-seater: Honda Insight, 61 mpg city, 68 mpg hwy

- Most efficient subcompact: VW New Beetle (diesel), 42 mpg city, 49 mpg hwy

- Most efficient compact: Toyota Prius, 52 mpg city, 45 mpg hwy

- Most efficient midsize: Honda Accord, 26 mpg city, 34 mpg hwy

- Most efficient large: Chevrolet Impala, 21 mpg city, 32 mpg hwy

- Most efficient small station wagon: Volkswagen Jetta Wagon (diesel), 42 mpg city, 50 mpg hwy

- Most efficient midsize station wagon: Ford Focus, 27mpg city, 36 mpg hwy

The least efficient cars

- Least efficient two-seater: Ferrari Enzo, 8 mpg city, 12 mpg hwy

- Least efficient subcompact: Ferrari 456 MGT/MGA, 10 mpg city, 15 mpg hwy

- Least efficient compact: Bentley Continental R, 11 mpg city, 16 mpg hwy

- Least efficient midsize: Bentley Arnage, 10 mpg city, 14 hwy

- Least efficient large: Bentley Arnage LWB, 10 mpg city, 14 hwy

- Least efficient small station wagon: BMW 540I Sport Wagon, 17 mpg city, 21 mpg hwy

- Least efficient midsize station wagon: Audi S6 Avant, 15 mpg city, 21 mpg hwy

What's Your Mileage?

What can you do? You can start by figuring out what kind of gas mileage your car gets. It's not that difficult. The next time you go to the gas pump, write down the mileage. You have to fill the car up or you can't figure the mileage. Then the *next* time you get gas, fill the car up again, and note how many gallons you used. Also note how many miles you drove, either by subtracting the old odometer reading from the new, or by using one of the alternate readings on the odometer. Divide the number of miles you drove by the gallons you used (what it took to fill it up again). The result is your mileage per gallon of gas.

Try it a few times. Note the difference in the mileage when you drive it around town and when you drive it on the highway. Pay attention to how your mileage goes down when you drive it over 60 mph (96 kph). If your car is in tune, you'll get better mileage and you will be dumping less carbon dioxide into the atmosphere. You can increase the fuel efficiency of your car by keeping your car maintained and by watching your driving habits.

Auto Maintenance

A well-maintained car will get better mileage and emit less greenhouse gas. You'll spend less at the pump and the atmosphere will be cleaner. Here are some tips:

- **Tune that engine** A finely tuned engine has more power, idles better, gets better mileage, and creates fewer emissions. Get your car tuned once a year. Follow the manufacturer's recommendations. Shop around for a good mechanic. The cheapest one isn't always the best one. Check your oxygen sensor regularly. Make sure the catalytic converter is working properly. Simply replacing the spark plugs can make a difference. A car that's out of tune, has a faulty or plugged catalytic converter, or a faulty oxygen sensor can reduce your mileage by as much as 40 percent. That means that you may be using 40 percent more gas than you need. That can mount up fast.

- **Check the air filter** Have your mechanic check your air filter and replace it regularly. An air filter catches the dirt and other impurities before it enters your engine. That dirt can wreck your engine as well as your mileage. Replacing a dirty filter can increase your mileage up to 10 percent.

- **Keep those tires inflated** This may be the easiest one of all, but it's important. Making sure your tires are properly inflated will increase your mileage, make your car handle better, and make it safer to drive. It's a good idea to use your own tire gauge to check your tires. The tire gauges on the air hoses at the gas station are notoriously inaccurate. Get a gauge at an auto parts store and check your tires before and after you fill them up.

- **Get the right oil** Use the manufacturer's recommended grade of oil. Using a 10W-30 oil in an engine designed to use 5W-30 can lower your mileage. Keep your oil clean. Manufacturers tend to give you a range of mileage recommendations for oil changes. Consider using the lower mileage to keep your engine running smoothly.

- **Watch that lead foot**   If you like to speed, accelerate rapidly, slam on the brakes, weave in and out of traffic, guess what? You're wasting gas. Not only are you a lot more apt to get into accidents, but your mileage will suffer in the process.

> **CAUTION**
>
> **Global Warnings**
>
> Gas mileage decreases rapidly as you get over 60 mph (96 kph). Each 5 mph (8 kph) you drive over 60 mph (96 kph) is like paying an extra 10 cents a gallon of gas. Not only is high-speed driving risky to you and the vehicles around you, it's expensive.

- **Don't idle**   When you idle your car, you are getting 0 mpg. The bigger the car, the more gas you waste when you idle. Don't leave your engine running unnecessarily. Driving cautiously for the first few minutes is a better way to warm up your car than to sit there idling.

- **Just cruise**   If you've got cruise control, use it. Driving at a constant speed saves gas. Use your overdrive gear on the highway. It will reduce the rpms (revolutions per minute) of your engine, reduce engine wear, and save gas.

- **Combine those trips**   Got errands? Well, do them all at once rather than separately. If you combine your stops, you'll save time and money. Several short trips taken from a cold start will use up to twice as much fuel as a longer multipurpose trip covering the same distance but with a warm engine. Your car is much more efficient when it's warmed up. Cold starts are expensive. Keep them to a minimum.

- **Try telecommuting commuting**   If you can go to and from work during the off hours, you'll avoid traffic, get there quicker, and use less fuel. Tell your boss it's a win-win situation. Ask him or her if you couldn't telecommute one or two days a week from your home. That will save a lot of mileage. If you own more than one vehicle, drive the one that gets the best mileage. Save the truck or SUV for times when you really need it.

> **Climatoids**
>
> Consider carpooling, even if it's just one day a week. Many urban areas allow vehicles with multiple passengers to ride in special high occupancy lanes. If every commuter car carried just one more passenger, we'd save 32 million gallons of gasoline every working day.

- **Take the train or walk**   Take the train or other public transit. You'll not only arrive more refreshed, but you can take the time to read the paper or do some work. Or consider walking or riding a bicycle. Just trimming five miles off your normal driving pattern can eliminate 2,000 pounds of carbon dioxide in the atmosphere each year.

◆ **Roof racks** Roof racks are great if you've got a small car and need to take a trip. They'll give you more cargo space, and you can enjoy the economy and efficiency of the smaller car in town. But be advised that loads carried up top increase the aerodynamic drag of the car and decrease your fuel efficiency. If you have the option, stick your stuff in the trunk; don't carry it up top.

Choose a More Efficient Vehicle

I recently purchased a four-wheel-drive pickup truck. It takes me places out in nature I need to get to that a passenger car won't go. But I got a four-cylinder rather than a six. I previously owned a six-cylinder car and, frankly, couldn't imagine getting a four-cylinder. The last four-cylinder I had was a Volkswagen Bus. I owned it for several years and had to replace the engine twice.

But I wanted to be ecological, so I went ahead and got a four-cylinder. I expected it to be under-powered, but it wasn't. Cars are far better engineered these days than back in the 1970s when I was driving my bus around. My car gets good mileage, is economical, and has plenty of power. I go up hills just fine. I might downshift, but I don't have to pull over to let the other cars pass me. If I enter the highway and need to get up to speed, there's plenty of it. Even loaded up with traveling and camping gear, it runs fine.

> **Warm Words**
>
> A **hybrid** vehicle is one that utilizes both an electrical motor and a gas-powered engine which work in tandem to substantially decrease fuel consumption.

I essentially have a light truck, the one we've been complaining about, but it *is* a four-cylinder. My wife, however, is looking at buying a *hybrid* car. My cousin and my sister-in-law both have them. They claim they have no problem on the highway. My sister-in-law brags how she drove from Los Angeles to San Francisco (425 miles—680 km) on 10 gallons (37 liters) of gas!

Hybrid Cars

Hybrid cars are the first alternative-powered car on the market that seem to be catching on. They save gas by combining an electric motor with a gasoline-powered one. Over 150,000 were sold in the first few years of their introduction. Toyota and Honda were the first on the scene with a hybrid. But now a number of American companies, afraid of ceding another fast-moving market to the Japanese, are coming out with their own. Hybrid SUVs might soon counter a little of the SUV's bad press.

The electrical motor in the hybrid gets its energy from batteries that are charged by the gasoline engine when it is running or by collecting energy from the car's brakes when they are applied. In the Honda Civic and the Insight, an electric motor assists when the car is climbing a hill or accelerating sharply. In the Toyota Prius, the electric motor takes over at low speeds. In both cars, the gas engine shuts off and the electric one takes over when the car is stopped.

The hybrid's rise has been encouraged by clamoring from environmentalists and by stricter environmental regulations in a number of states, most notably California. That state mandated that a certain percentage of these cars be made available if car makers wanted to do business there. California is one of the biggest car markets in the world and also suffers some of the nation's worst pollution.

Sport utility hybrids are in the works that will soon start getting 40 mpg (16.6 kpl) or better compared to recent averages closer to 20 mpg (8.4 kpl). Though these cars currently cost about $5,000 more than other models without the hybrid function, that's expected to drop as production increases. Even now, if you factor in state and federal tax incentives to buy the car with the gas savings over the life of the car, you'll come out even. And you won't be polluting the atmosphere nearly as much. As more cars are sold, prices will come down, particularly as gas prices continue to rise.

Climatoid

Even the U.S. Army has been jumping on the hybrid bandwagon, developing a hulking hybrid battle vehicle, built by GM. Hybrid Humvees might soon join the circle. The military likes the improved mileage but also appreciates the fact that in the electric mode the car runs silently, making it great for reconnaissance missions.

Electric Cars

Under California's clean air rules, 10 percent of the vehicles sold in the 2003 to 2008 model years must be electric or zero-emission vehicles. But the state, realizing that the car companies weren't on track to reach that goal, offered to allow them to sell hybrid vehicles.

If we are to stop emitting greenhouse gases into the environment, some radical new technologies must be tried. The electric car was one vehicle that held high promise in the '80s and '90s, but the promise hasn't materialized. Electric cars have limited ranges, and they take considerable time to recharge. Rapid recharging technologies have not been developed.

In the early years of electric car development, some thought there might one day be a strip in the road that autos could recharge on, much like how streetcars once got energy from overhead cables. But that hasn't worked out. Using an electric vehicle for short runs, however, still makes sense. My uncle lives in a retirement community where a lot of people drive rechargeable golf carts to run errands. Some communities even have separate lanes for these cars.

Fuel Cells

There has been a lot of talk about fuel cells. Unlike batteries that store energy, fuel cells produce electrical energy by combining atmospheric oxygen with hydrogen or hydrocarbon fuels. Fuel cells need not be recharged. Most automakers are positive about the future of fuel cells, but they look a long way off. There are no fuel cell vehicles in production at present and it's going to be some years before we see any on the highway. Still they hold some of the brightest promise, since they could reduce our dependence on fossil fuels altogether.

Is Cheap Gas a Good Deal?

In Europe they pay a lot more for gas than we do in North America. Taxes are substantial on gas in the United States, but they're minimal compared to Europe where the cost of the gasoline actually accounts for only about 20 percent of the price of the fuel. In Europe, it's common to pay $3.00 to $4.00 a gallon ($.78 to $1.04 a liter). This sounds like a lot to Americans, but high petrol costs make *sure* that the consumer is concerned about fuel efficiency. Small economy cars and lots of motorcycles are common in Europe. Gas costs too much to be wasteful.

What about trying something like that in North America? Of course, European cities are more compact. You often need to drive a lot further to find a grocery store in America, particularly in the west. If we raised gasoline prices, though, people would become more cost-conscious. I'm not talking about increasing taxes; I'm saying shift more of the tax burden to gasoline consumption. Make gas taxes pay for everything there is to do with government support of transportation. Our gasoline taxes would pay for roads, pollution control, police services, street cleaning—anything to do with transportation.

The trouble is … politics. Have you ever noticed how gas costs drop in North America immediately preceding an election? People don't like it when gasoline costs go up. They think somebody isn't doing their job. Maybe it's time to try somebody else. Politicians pay attention to stuff like that. The public needs to accept that there is a price to be paid in terms of the environment for the privilege of cheap gas. If we paid more at the pump, our oil reserves might last a little longer.

Got Oil?

A lot of people have been saying for a long time that we are going to run out of oil some day. We've heard it so much, perhaps, that some think it's never going to happen. There will always be more oil, and more major oil reserves. It's a huge planet, there's a lot of black gold down there, and we just need to learn how to get at it. Maybe dig in a couple more national parks. Do a little more deep-ocean drilling.

But the truth is, oil can't last forever. The 100-year petroleum era may be coming to an end. M. King Hubbert, a Shell geologist, predicted in 1956 that U.S. oil production would peak in the early 1970s. Some people thought he was God, some thought he was a jerk. But in 1970, when U.S. oil production did peak and start its long decline, a lot of the skeptics stopped being skeptical.

In the book, *Hubbert's Peak: The Impending World Oil Shortage*, author Kenneth S. Deffeyes uses Hubbert's theories on worldwide oil production to predict that somewhere between 2004 and 2008, world production will peak and start to fall. There are a few things that could spoil Deffeyes's predictions. We could find huge new oil reserves. We could also develop new oil drilling processes that could wrest more oil from known reserves. Or oil prices could climb so high that it would become profitable to squeeze the last drops out of some of the most stubborn reserves.

Finding new oil reserves is unlikely. Petroleum geologists have been everywhere, and there have been no new major finds since the 1970s. Billions have been poured into drilling technology and it just may be as advanced as it's going to get. And lastly, even high oil prices won't avoid the inevitable. Europe has high prices, and they are still hungry for oil.

According to Deffeyes, the bind will come in the next decade as we move away from our dependence on crude oil. He claims we must move as quickly as possible to natural gas, nuclear power, solar, wind, and geothermal energy.

Hot Debate

According to Kenneth S. Deffeyes, a geologist who once worked for Shell Oil and who now is a professor at Princeton University, global oil production will peak between 2004 and 2008 and begin its slide downward. In his book, *Hubbert's Peak*, he states, "Fossil fuels are a one-time gift that lifted us up from substance agriculture and eventually should lead us to a future based on renewable resources."

The Least You Need to Know

- Raising the fuel-efficiency standards of our vehicles would be the most effective thing we could do to stop global warming.

- Light trucks and SUVs get around fuel-efficiency standards established in the United States in the 1970s because the law considers them work vehicles.

- Little work is done in these vehicles; they need to be held to national standards.

- Driving sensibly and keeping your car in tune will help our air.

- Hybrid cars are the most fuel-efficient promise presently on the market.

- Cheap gas might be a luxury the environment can't afford.

23

The Rainforest Is on Fire

In This Chapter

- ◆ Treasure in the tropical rainforest
- ◆ Involving the natives
- ◆ Ecotourism
- ◆ The jungle needs help

The tropical rainforests of the world are a unique treasure. Rainforests are found around the world, mostly near the equator. They support an incredible array of wildlife. Scientists estimate that anywhere from 50 to 90 percent of all animal species are found in tropical rainforests, and many of these species haven't been discovered. But rainforests are under attack from farmers, ranchers, miners, oil companies, and developers trying to clear the land. Satellite images of the Amazon jungle show more than 8,000 fires burning in a single day.

As the trees are cut and burned, they release a tremendous amount of carbon dioxide into the atmosphere. The destruction of the forest and the tilling of the land produce about 25 percent of the greenhouse gasses released by man into the atmosphere. The tropical rainforest needs some friends. Let's hold a concert.

Concert in the Jungle

I am sitting in the ruins of the Plaza Mayor, the ceremonial gathering point of the 1400-year-old Mayan City of Tikal in Guatemala, waiting for the concert to begin. It's 5:30 A.M. and the sunlight is just starting to filter through the thick jungle canopy that envelops the ruins.

The symphony begins with the flutelike songs of the melodious black birds, followed by a chorus of high caws from the Aztec parakeets, and the red lored parrots. The tempo picks up with the hysterical humanlike cackles of the laughing falcons as the lineated woodpeckers pound out the backbeat on tall swaying palms. The crescendo continues to rise until daybreak when the howler monkeys join their avian brothers and sisters with a cacophony of lionlike roars that can be heard for miles.

Tikal is to Guatemala what Banff National Park is to Canada, or Yellowstone is to the United States. Protected since 1955, the wildlife within the park's boundaries is both plentiful and oblivious to human intruders. Tikal is located in the subtropical forest of the state of El Petén in northern Guatemala. Within the boundaries of the park, the forest is thick and virgin—almost completely undisturbed.

In populated areas just outside the park, however, the forest makes up only a quarter of the terrain, and very little of that is undisturbed. Guatemala is a microcosm of what is happening in other tropical third world nations in Latin America, Africa, and southeast Asia. Virgin rainforest originally covered 4.2 million square miles (11 million square kilometers) of the tropics, but now only 380,000 square miles (1 million square kilometers) remain.

Global Warnings

Scientists believe that the collapse of the Maya might have been due to climate changes, as well as the mismanagement of their forests. As forests disappear, so does wildlife and local plant species. The collapse of the Mayas might be one of history's first ecological catastrophes.

Lessons from the Maya

Tikal was a thriving center of Mayan commerce. At its height between 700 and 800 C.E., this citadel once held a population of over 90,000 Mayan Indians with over 4,000 structures, including plazas, temples, palaces, ball courts, and pyramids. A number of these monumental structures have been excavated, though most are still buried beneath the dense green jungle. The collapse of the Maya, scientists speculate, is attributable to both drought and deforestation, the latter of which led to a collapse of the natural food chain that surrounded them.

The Rainforest as a Sink

The rainforest is a vital component in the take-up and release of carbon dioxide in the atmosphere. Trees and other plants absorb carbon dioxide while growing, and they release carbon dioxide when they decay. They also release carbon dioxide when they are burned. The reason the tropical forest is so important to the atmosphere is because of its density. In tropical rainforests you rarely see the bark on trees. It's covered with moss and other plants. It's hard to see the animals because the growth is so thick.

This is why it's such a good idea to save it. The density of the forest creates an enormous sponge that can forgive a lot of our industrial sins by sucking up carbon dioxide from the atmosphere. The Kyoto agreement allowed industrial nations to receive credits toward their reductions of greenhouse gases by replanting tropical forest.

Regrowing temperate forest is likewise a good idea. Spain has done a good job of that. But temperate forest can't match the density of plant life and wildlife that is in the tropical rainforest nor can the temperate forest match the rainforest's carbon absorption ability. And saving the tropical rainforest now is a lot easier than replanting it later.

> **Climatoids** _____
>
> A few countries, such as Spain, have demonstrated that temperate forests can be restored to the land even after centuries of abuse. Spain has reforested nearly 5 million acres (2 million hectares) since the end of World War II.

Slash and Burn

Tikal is located in the northern lowlands of Guatemala, in the Province of Petén, which is typical of agriculturally developed tropical jungles. Though the jungle (or tropical rainforest as it is variously referred to) is a vital and rich biological system, the soils are as good as sterile. "You scrape away the leaf littler," says Dave Whiticre, head of the Maya Project that is studying the raptor community in the Peten for the Peregrine Fund, "and you'll see there is a thick layer of leaf litter and essentially no humus. You get down to mineral soil almost immediately."

The leaf litter is rapidly decomposed by the incredible mass of jungle insects. Leaf-cutter ants cut paths 6 or 7 inches (15 to 18 centimeters) through the leaf litter in Tikal. And the heavy rains leach nitrogen and potassium out of the soil. The nutrients in the tropics are essentially held within the tree, so when the forest is destroyed, the land is poor for agriculture.

On the road from the Flores airport in Peten to Tikal, Whiticre and I stop the car at a section of patchwork jungle that exposes the series of successions that typically occur with the slash-and-burn method of agriculture. Here farmers take a tract of forest and cut it down during the dry season. They let the wood dry for a month or so and then they torch it before the rainy season. Thus the nutrients in the trees are released in the form of ash. The farmer tills the soil and plants his crop just before the rains.

During the first season the crop is good, the insect pests and weeds are at a minimum, and the soil is rich, but with each successive crop, the soil gets poorer, the insects and the weeds build up, and the output grows weaker. In some areas, farmers can get five to ten crop rotations, but in most areas, after two or three rotations the field is abandoned and the farmer moves on to cut another section of the jungle.

> **CAUTION**
>
> ### Global Warnings
>
> Deforestation in the tropics can change the weather hundreds of kilometers away. Scientists have found that reduction of the lowland forest affects the amount of moisture which rises with the clouds to the higher mountain jungles. The resultant reduced cloud cover is affecting the high jungles in Costa Rica, extending the dry season and diminishing amphibian populations.

Spreading the Message

The Peregrine Fund, a conservation group out of Boise, Idaho, dedicated to the preservation of the world's falcons, hawks, and eagles, has just wrapped up a long-term study of the park's raptors in which they schooled 115 Guatemalans in forestry management and wildlife biology. These people were used by the Peregrine Fund to monitor the behavior of a number of the park's raptors—including laughing falcons, barred forest falcons, and ornate hawk-eagles. The conservation group's aim was to teach a respect for the birds and the land to their charges in hopes they would take the message home and it would spread throughout the region.

The advantage of getting the locals interested in top predators like these raptors is that it takes a lot of jungle to support top predators. And within that territory a lot of smaller species can flourish. Also, hawks, eagles, and falcons as well as jaguars, ocelots, and other predators appeal to the Latin macho ethic, giving even tough young males an excuse to fall in love with nature.

Tracking wildlife in Guatemala.

(Photo by Michael Tennesen)

Conservation Approaches

Conserving the tropical rainforest requires a variety of approaches. Conservationists need to provide economic alternatives to local populations so natives can make a living doing something besides cutting down the trees. Conservationists need to provide education so locals can learn the value of the rainforest and the species contained. And conservationists need to work with international corporations to make sure business doesn't contribute to further *biodiversity* loss.

> **Warm Words** _____
>
> **Biodiversity** describes the number of species in a given environment. It is shorthand for "biological diversity."

Alternative Rainforest Products

Forest communities create enterprises from oil, nuts, fruit, plants, fibers, and other natural materials sustainably harvested from trees and plants, thereby creating an incentive to keep the forest standing. In Peten, a number of locals harvest chicle from the sapodilla tree, an evergreen that flourishes in the rain forest. Guatemalans harvest the sap without cutting down the trees. The Mayan Indians chewed chicle more than a thousand years ago. Chewing gum made with chicle was developed in the 1860s.

Much of the coffee and cocoa produced in this region is grown on plantations where the forest was cut back to encourage the growth of the beans. But many farmers still produce shade-grown coffee and cocoa which has become popular in the United

States in organic markets. Starbucks even has shade-grown coffee. By choosing shade-grown (organic) coffee or cocoa, North Americans can be sure that all of the trees weren't cut down to make the coffee or cocoa beans grow faster. These trees are vital to wildlife and the uptake of carbon dioxide.

Historically, all coffee was cultivated in the shade beneath a forested canopy. But in the early 1970s, fostered by grants from the U.S. government, farmers in several countries began growing varieties of coffee that thrive in direct sunlight. The goal: to increase production. But without the natural protection provided by a surrounding forest, sun-grown coffee usually requires heavy doses of pesticides and chemical fertilizers. Researchers at Conservation International found that, in some places, growers use about a pound of toxic chemicals for every pound of sun-grown coffee they produce, making it one of the most chemically intensive crops on the planet.

Climatoids

You can encourage the saving of the rainforest by the kind of coffee you drink. Shade-grown coffee (most organic coffee is shade-grown) is the one to ask for. Ninety percent of the coffee purchased in America is sun-grown, which means that the rainforest was cut down to maximize growth. By asking for shade-grown coffee or organic coffee the next time you're at your local coffee shop or grocery store you'll be making sure that major portions of the rainforest remain intact.

Butterfly Farms

One successful alternative enterprise practiced in Africa is the butterfly farm. In Kenya, on the east coast of Africa, there are more than 500 butterfly breeders, part of a United Nations project named Kipepeo, which means "butterfly" in Swahili. Butterfly farmers adjoining the Arabuko-Sokoke forest are a buffer to woodcutters and developers. These farmers attract gravid butterflies to their properties and then harvest a portion of their eggs. The eggs change into caterpillars and then into chrysalises (cocoons), which are shipped to butterfly display houses in America and Europe. There the chrysalises turn into butterflies.

Some of these farmers were once poachers that ravaged the forest for wood and wildlife, but butterfly farming is more lucrative than poaching. Now when it comes to preserving the coastal African forest, the butterfly farmers are the most vocal group around.

Ecotourism

Ecotourism is another way we can help the tropical rainforest. The term ecotourism began as little more than advertising copy for high-priced adventure travel in the 1970s and 1980s. But since 1990 it has become a buzzword in both the travel industry and the environmental movement. It really has two meanings. In the crowded natural parks and wilderness areas in the United States and Canada, ecotourism refers to minimizing visitor impacts on our parks and wild places.

Warm Words

Ecotourism is environmentally friendly tourism. In North America it refers mostly to minimizing visitor impact. In third world nations it more often refers to supporting local economic alternatives to depleting natural habitats.

At Home

Ecotourists in the Everglades put canoes into a special section of the park, known as the Nightmare, where motor boats are not allowed. The Nightmare is a narrow section of the swamp, overhung by mangroves, where visitors can see egrets, herons, roseate spoonbills, and osprey. Alligators occasionally jump into the water as you pass.

Hot Debates

Managing tourism's impact in natural environments sometimes requires methods which some argue defeat the purpose of pristine areas. One example is Isle Royale National Park, which is 90 percent wilderness. In this forested wilderness where the "hand of man" is not supposed to show, the Park Service has built boardwalks in many of the boggy areas. Park rangers point out that when people walk through a boggy area, the mosses break down under foot, creating muddy trails, which gradually widen with continual traffic. With a wooden path, however, the damage is constrained.

The tent cottages at Maho Bay Camps in the Virgin Islands National Park on St. John have won numerous awards for blending into their tropical background and for the camps' environmentally friendly approach. The tent cottages, built on 16-square-foot platforms, are made of translucent fabric on wooden frames, with screened terraces to catch the cooling trade winds. The facilities utilize solar and wind energy, composting toilets, and energy-efficient appliances. Old newspaper is the only protective wrapping used at the Maho Bay Store, and the camps' bar gives a 25-cent discount if guests will reuse the same mug on refills of beer.

To stay at such an ecohotel, check out www.greenhotels.com, which lists such establishments from Philadelphia to Hawaii. An example is the Summit Views Lodge in Los Gatos, California, which uses Goodwood Firelogs made of compressed wood chips and sawdust held together by all-soy wax. The log is delivered to the room in a food-grade jute bag, which can be set directly on the fireplace grate. It's eco-friendly, because it uses existing byproducts rather than cutting down more trees. Plus, there is no waste packaging left over.

Abroad

The trend toward ecotourism in *international* parks and wildernesses is a bit different. Ecotourism in foreign countries has centered more on support for local communities so that locals will have a reason not to exploit timber, wildlife, and other natural resources in protected areas.

Vendors displaying beautiful textiles and wooden carvings are located along the road up to Tikal National Park. By hiring a local guide and by purchasing native crafts you are giving them an alternate source of income to cutting and burning the forest to raise corn. Conservationists believe that most Latin American Parks are not endangered by too much tourism. With the funding problems that most protected areas have in Latin America, what is most needed is a heck of a lot more tourists.

Getting the Natives Involved

A few years ago I traveled to Vilcabamba in the Peruvian rainforest with an international team of biologists sponsored by Conservation International. They were trying to assess the species diversity of this high mountainous region between the Andes and the Amazon for consideration as a park. I got to witness how the biologists worked to involve locals in their work.

Early one morning I accompanied Tom Schulenberg, an ornithologist at Chicago's Field Museum and leader of the expedition. With us was Lawrence Lopez, a Peruvian ornithologist, whom Schulenberg was trying to train. The biologists worked around the edge of the bog that centered our study site, training both binoculars and microphones on the surrounding jungle. According to Schulenberg you can hear approximately three to four times more birds than you can see, and he was anxious to teach this ability to Lopez. All American biologists on the expedition had Peruvian counterparts they were trying to train.

Teaching ecology to the locals.

(Photo by Michael Tennesen)

The population of Peru is gradually becoming more environmentally aware. Monica Roma got her masters degree in ecology at the University of Missouri and remembers how in 1995 when she returned to Peru, ecology was the fashion. Five years earlier no one even knew the word. "But when I got back, people were protecting the dolphins, recycling their garbage, and talking about saving the rainforest."

Our Way

International conservationist groups encourage the World Bank and the Inter-American Development Bank to emphasize environmental needs when loans are made. Those development loans channel millions of dollars into poor countries for road projects, dam building, creation of export-agricultural plantations and other useful efforts. But poorly planned projects funded by multilateral lenders have resulted in human colonization of rainforests and in agricultural projects that weren't always in the best interests of the rainforest.

The problem is that the United States and international development organizations often encourage their own economic models—the export of agricultural commodities. But this may be one of the root causes of deforestation. And deforestation often leads to hunger because of the limited amount of agricultural products the natives can get from cleared jungle before the land turns barren.

Forgiving Debts

One innovative strategy is debt-for-nature swaps. In these agreements, an environmental organization buys up a portion of a developing country's international debt; in return, the country's government agrees to put in effect local conservation programs or to preserve wilderness. The agreements provide some relief from the country's share of an enormous third world headache: the more than one trillion dollars the world's poorer nations owe banks and development agencies. Servicing that debt has led cash-starved governments to cut funds for environmental programs, and encouraged them to plunder resources.

Preserving the Rainforest

Grabbing up much of the tropical rainforest and preserving it as parks and wilderness areas is another idea. Costa Rica has preserved much of its rainforests in national parks and enjoys a fair amount of international ecotourism as a result. I visited the cloud forests of Monte Verde in that country and spent a pleasant stay in a local lodge with an Australian, a group of Brits, a couple from France, and several Germans.

Wilderness Parks

One of Conservation International's principal strategies focuses on major tropical wilderness areas. The three principal areas they promote are Amazonia in South America, the Congo Forest in Africa, and New Guinea in Indonesia. These areas are the largest remaining tracts of pristine tropical forest on Earth and are more than 70 percent intact. Tropical wilderness areas are among the last places where indigenous people can maintain traditional lifestyles. And they are of crucial importance to climate regulation.

Roads

Part of the urgency that biologists gave to the expedition into Vilcabamba was that oil giants Shell, Mobil, and Chevron had won concessions from the Peruvian government to explore the valleys east of Vilcabamba, extending to Bolivia. If oil or natural gas was to be found, a pipeline might be built over the Andes.

Things like pipelines, oil wells, and commercial logging make biologists and conservationists nervous. All of these activities introduce roads that provide access to the deep forest. That access brings illegal hunting, indigenous farms, and population growth.

> **Global Warnings**
>
> Conservationists are often less afraid of the oil exploration, mining, and commercial logging operations in the tropical rainforest than they are of the roads these enterprises bring. After the oil wells are gone, the roads they leave behind provide access to formerly inaccessible regions of the forest, bringing poaching, slash-and-burn agriculture, ranches, villages, and the eventual demise of the tropical rainforest.

Hot-Spots

Besides promoting wilderness, Conservation International also promotes what they call hot-spots. Hot-spot is a term coined to identify areas with large numbers of species that are concentrated in a relatively small space. If we can save these areas, we will be able to preserve a maximum number of species for the limited amount of conservation dollars available.

Conservation International lists some 25 areas as hot-spots. Those include the tropical Andes, Madagascar, South Africa's Cape floristic region, the Antilles, the Philippines, Brazil's Atlantic forest region, the Polynesian and Micronesian island complex including Hawaii, southwestern Australia, the Mediterranean region, the Guinean forests of west Africa, and others.

> **Global Warnings**
>
> Hot-spots have the highest concentrations of unique biodiversity on the planet. They are also the places at the greatest risk of destruction. The need for conservation in these regions is urgent to prevent a wave of species extinctions.

The 25 hot-spots identified by Conservation International contain 44 percent of all plant species and 35 percent of all terrestrial vertebrate species in only 1.4 percent of the planet's land area.

Diversity

Vilcabamba is in the middle of the tropical Andes and is incredibly diverse. Many biologists believe that the tropical Andes are the most biologically diverse of all the remaining forests in the world. The Vilcabamba Range, roughly the size of New Hampshire, is an uninhabited cloud forest in the midst of this towering wilderness. Cut off from the nearby landmasses by the deep valleys of the Apurimac and Urubamba Rivers, it rises like an island in a sea of jungle and is as isolated as an island in the ocean.

Animals in such remote landscapes are isolated, too. Separated from the rest of the world, many populations have evolved alone and out of touch with each other over the eons. The result: pockets of diversity, often teeming with species never before described by scientists.

Climatoids

Biologists investigating the Vilcabamba Range, an uninhabited cloud forest roughly the size of New Hampshire between the Andes and the Amazon in eastern Peru, estimate that the area has at least 650 species of birds. That number is roughly equivalent to the total number of species that breed in the United States and Canada.

Their study sites demonstrated the incredible diversity of the land. Though its elevation was the upper limit for monkeys, there were white-faced capuchin monkeys, spider monkeys, and night monkeys. There were signs of puma, dwarf deer, and bear.

The mammalogists on the team collected a weasel, a rare short-tailed opossum that was previously represented by only two Peruvian specimens, and a rare bamboo rat that had hands like a monkey—a first for both the Peruvian and Smithsonian collections. They also encountered 16 species of reptiles and amphibians with as many as 12 being new to science. The biologists catalogued 142 species of birds. Schulenberg estimated that Vilcabamba had at least 650 species of birds.

One day I accompanied Brad Boyle, a Canadian botanist who specialized in South American plants, and his Peruvian counterparts, two young Peruvian biologists. We traversed the bog to the forest beyond, then climbed to a rise, where they laid out a 50-yard line and took specimens of the plant life within two to four yards of either side of the line.

That day their attention was directed upward at the epiphytes—bromeliads, orchids, ferns, mosses, and other plants—that crowded the limbs of the trees above. The enormous presence of these epiphytes is one of the factors that distinguish cloud forest from other types of jungle. Boyle pointed to a cluster of treetops in one area and declared that there were more species in that cluster than in most northern deciduous forests.

Cloud forest animals often specialize on individual plants or groups of plants. Some flowers have long curved tubes that only cycle-billed or sword-billed hummingbirds can pollinate. But there are also cheaters, like flower piercers, that have hooked bills like a can opener that can make little holes at the base of the flower, and there are cheater bees and other smaller hummingbirds that use those holes to get at the flower's nectar.

Boyle expected to find many new species in such an isolated forest, but said describing new species was a long and arduous task. "It's not like 'Eureka! I just found a new species,' it's more like 'Wow, I've just been through every known species in the herbarium and it doesn't match anything.'" Still he held up a tiny delicate orchid and dared to declare, "I'll bet you a case of beer that's never been described before."

Specialists

Plants and animals are distributed differentially on the globe. Some, like humans and gray rats, are widespread and inhabit most regions, while others have very restricted distributions and may be found only on a single island, in a single river or lake, or on a single mountaintop. We refer to these highly restricted organisms as *endemic* species because they are unique to a specific region. Many of the species that occurred in the Vilcabamba range were endemic, found nowhere else in the world.

Warm Words

Endemic species of plants or animals are those that are found only in restricted areas—a single island, a single river or lake, a single mountain range, or a single mountaintop.

Ants Won't Stop Progress

The areas where endemic species flourish are often difficult, far removed, and treacherous. Vilcabamba was one of these difficult places. When I exited the helicopter, my leather boots sank six to eight inches into a bog. They stayed wet the entire two weeks I was there. The earth, saturated with water, was covered with a deep, spongy matrix of moss. Here and there were sinkholes that could swallow your leg to the thigh.

The team cleared paths with machetes. The branches that remained after the machetes were sharp and acted as natural booby traps if you walked too rapidly. The trails were like obstacle courses with stumps and tangled roots. I learned to walk on all fours. One of the Smithsonian biologists, Louise Emmons, slipped on the trail and fell into one of these cut branches. Though there was no x-ray machine in camp, she felt she'd cracked her rib. The worst part was that she'd been nursing a bronchial infection for a number of days and now couldn't cough.

In the evening I thought about taking a bath but Brad Boyle warned me, "The army ants are swarming down by the river. Be careful. You don't want to get one up your pant leg." I decided to wash my clothes instead. Wool socks took two days to dry. My cotton T-shirts never dried.

At night I thought of the beauty of the cloud forest—the orchids, the mosses, the ferns—but I also thought of the hardships. I wondered if Vilcabamba was tough enough to save itself. I proposed this to Emmons, but she disagreed. "Too many times we've assumed that places were indestructible or too remote and we come back ten years later and they're gone."

If the world wants to have the tropical rainforest in its future, we need to start saving now.

The Least You Need to Know

- The destruction of the forest and the tilling of the land creates about 25 percent of the greenhouse gasses released by man.

- Tropical rainforests are the greatest storehouse of species diversity and density on the planet.

- The rainforest is also a vital component in the take-up of CO_2 from the atmosphere.

- Agriculture, mining, oil exploration, and the roads they bring all pose serious threats to the continued existence of the rain forest.

- We need to provide alternative methods of commerce to native populations if we are going to keep them from cutting the rainforest down.

- The difficulty, impenetrability, and pure hostility of some rainforest areas are not enough to hold back the tide of development.

Running Out of Gas

In This Chapter

- Bustling futures in clean fuel technologies
- Solar, wind, and other alternative sources
- Clean nuclear energy
- Everyone's wild about hydrogen

Though electricity was used in the early 1800s, it wasn't until Thomas Edison opened the first electricity-generating plant in London in 1882 that the energy age really got going. Cities got electricity first. Rural areas in the United States didn't get theirs until the 1930s, but it's been growing ever since. The U.S. population grew 79 percent from 1949 to 1997, but electric usage grew by over 1100 percent.

And we've been looking everywhere for coal, oil, and natural gas ever since. About 80 percent of the oil produced today flows from fields that were found before 1973, and the great majority of them are declining. New exploration tools haven't improved the oil discovery rate, even though today's tools and techniques are much improved from earlier decades.

U.S. oil production peaked in the 1970s and has been going down ever since. Many scientists believe the world's production will follow the same production peak and decline in the next ten years. We will experience severe shortages of natural gas first, followed by oil, followed by coal.

The problem is our cleanest energy is natural gas, what we're trying to convert to. The dirtiest is coal. China has the greatest reserves of coal, but if they end up using it to fuel accelerating development, it could neutralize Western efforts to reduce pollution and greenhouse gases.

Global Warnings

Crude oil reserves in the United States are estimated at less than 21 billion barrels, just 2 percent of the world's reserves. Central and South America have 6 percent, Africa 8 percent, Russia 7 percent, Canada and Mexico 5 percent combined. The Middle East has 64 percent of the world's reserves. If you don't think that's a problem, then you haven't been reading the newspapers.

But don't despair. A dose of global warming may bring on depression, but the antidote could be the best thing that ever happened. Opportunity is knocking. Fossil fuels have a diminishing future. Alternative energies have an unlimited one. It will be hard, but we can do it. We went to the moon. This is a piece of cake. Besides, there's money to be made.

What We've Got

Though it may take time to develop a full suite of energy alternatives, there are things we can do right now to clean up the energy we've got.

Coal

Coal produced 75 percent of our energy by the end of World War I, but then oil moved in and began to fuel our trucks, cars, and locomotives. Oil was cleaner and easier to move around. Natural gas muscled in and took over the cooking and heating market. But coal moved back into dominance in the late 1970s when it became the principal fuel for electric utilities.

The trouble is coal is about 70 percent carbon and produces more carbon dioxide than oil or natural gas. It also emits more sulfur dioxide and nitrogen oxides. In the

United States the recent trend has been for electric utilities to move away from coal toward natural gas. Similar changes in Britain have radically reduced their greenhouse emissions.

It's possible to make coal cleaner. There are techniques that can remove up to 90 percent of the sulfur and ash. Those techniques include using microwaves, microbes, and enzymes. Coal gas has promise as a cleaner-burning fuel. Mixing finely ground coal with oil or water to produce liquid can reduce pollutants.

Climatoids

It takes three tons of coal for an electric plant to produce enough energy to power the average U.S. home for a year.

Oil

We've been using oil for 5,000 years to light our lamps, lubricate our machinery, seal our ships, make roads and even weapons. Ancient Arabs made a kind of napalm out of it. But it was the invention of the internal combustion engine that really turned us into an oil economy.

The trouble is burning oil creates pollution. It can be made cleaner, though. Reformulated gasoline, which is meant to reduce smog, is increasingly used across North America, particularly in California and the Northeast.

Climatoids

Oil accounts for 41 percent of the total U.S. energy package. Coal accounts for 24 percent and natural gas 23 percent.

Natural Gas

Natural gas is mostly methane, with ethane, propane, and a few other hydrocarbons mixed in. Methane, as you recall, is one of the greenhouse gases, though it has a shorter life in the atmosphere than carbon dioxide. When natural gas is burned, it's cleaner than any of the other fossil fuels. This is why the Clean Air Act Amendments of 1990 encouraged increased use of natural gas because of its low-combustion emissions.

The Hoover Dam.

(Photo by Michael Tennesen)

Hydroelectric Power

Hydroelectric power is both cheap and clean. It's also renewable, because the water that drives the generators comes back each rainy season. Hydroelectric power normally comes from dams, which create reservoirs. As the water goes through the dam, the force is used to turn generators creating electricity. Though hydroelectric power is gentle to the atmosphere, it's tough on the land, fish, aquatic species, Native Americans, and fishermen that utilize the river.

Warm Words

Hydroelectric power or electricity is created by the force of water driving a generator.

Dams can stop fish migrations, wreck spawning grounds, and slow the downstream flow of water, which impacts wildlife. Though fish ladders allow most of the fish to circumvent the dams, the environmental impacts of these structures are considered serious enough that few new projects are in the planning in North America.

Hot Debates

The Endangered Species Act of 1973 has halted further hydroelectric growth on the Columbia River basin, which has more than a third of the nation's hydroelectric capacity. Enforcement of the act has caused dam operators to release more water for salmon and other wildlife and has blocked new hydroelectric development.

Nuclear Power

From a global warming point of view, nuclear power has certain advantages. It emits few greenhouse gasses. France obtains 75 percent of its electricity from nuclear power. Other countries that rely on its use have fewer problems meeting Kyoto emission agreements than the United States. But alas, nuclear power has lots of bad points as well—like the Chernobyl and Three Mile Island accidents. North America and Europe have stopped building new plants and have turned an eye toward other nations who use nuclear fuels out of worry that nuclear bombs might be their intent.

Fission

Current nuclear power plants utilize fission to make energy. Heavy elements such as uranium are bombarded with neutrons, which break the uranium atom and release more neutrons. The process becomes self-sustaining and each time an atom is broken it releases energy. That energy is used to heat water, which drives a turbine that produces power. Each reactor core contains several thousand fuel assemblies composed of sealed fuel rods filled with uranium oxide pellets. However, over time these rods lose their heating ability and the assemblies have to be refueled. What to do with the spent fuel is part of the reason that nuclear power is on the wane in most western nations.

Fusion

But then there is fusion. Fusion is what the sun runs on. Fusion occurs on the sun when hydrogen combines to form helium or other heavier elements. It's a hot combination, releasing about a million times more heat than burning oil. The problem is hydrogen nuclei normally repulse each other, so you need a lot of energy to bring them together. Right now it takes more energy to cause fusion than you get out of the process. We've learned how to induce fusion in the hydrogen bomb, but you can't plug your stereo into it. But there are projects in the works, particularly in Japan, that could get this going.

The Promise of Solar Energy

The promise of solar power is more immediate. Solar power has come a long way since its early days. Vast areas in countries around the world that don't have power lines have fueled the growth. Solar energy comes in a variety of forms: solar water heating, passive solar heating and cooling, photovoltaic technology, and solar thermal technology.

Solar water heating uses a solar collector to absorb the sun's energy and then heat the water in a solar tank. A black 50-gallon drum filled with water set up on the roof of the house is the preferred solar water heater in Mexico's Baja Desert. Fancier devices work in higher latitudes and can deliver a surprising amount of hot water with no carbon dioxide or other pollution created.

Solar lessons from the natives.

(Photo by Michael Tennesen)

Passive solar heating works with a building's design to heat the house in the winter and cool it in the summer. Adobe homes in New Mexico lean on early Native American designs where windows face south to capture the sun's low rays in the winter, but are shaded by an awning to prevent those rays from entering the house in summer when the sun hangs higher in the sky. Landscaping that uses deciduous trees, which provide shade in the summer but lose their leaves in winter to allow in sun, follow the same principle.

Photovoltaic Cells

Photovoltaic (PV) cells hold the greatest promise for solar energy. They take in light from one side of the cell and send electricity out the other. It's as simple as that. Yet developing and improving these cells has been a monumental task. PV cells are created from various forms of silicon as well as film materials, some thinner than the human hair.

PV cells, however, create direct current, and since U.S. circuits run on alternating current, an inverter is needed. Most systems need switches to tie the PV system to the power grid, so the grid can act as a backup. In some areas it's possible to run excess energy into the power grid and get credit for the contribution. A PV solar system also needs batteries to store the energy. This provides the ability to charge up the batteries during the day and to power the lights at night. Fluorescent lighting helps these charges last longer.

Climatoids

To supply the United States with its current energy budget would take an array of photovoltaic cells covering 10,000 square miles (26,000 square kilometers). The total amount of photovoltaic cells shipped from 1982 to 1998 would cover a little over 1 square mile (3 square kilometers). A massive, but not insurmountable, scale-up is required to provide current U.S. energy requirements, and to gear up for the 2050s, when energy requirements may triple.

Solar energy can be made more efficient by concentrating sunlight into the cell. It's better used in areas that are mostly cloud-free. This requires a tracking device to align the sun to lenses or reflectors, which concentrate sunlight into PV cells.

The Oak Ridge National Laboratory is developing a full-spectrum solar energy system that will utilize not only the visible but the infrared portion of the spectrum of light. The system will actually capture both types of light, directing the visible spectrum via large-diameter optical fibers into building interiors for illumination while using the infrared light to power photovoltaic cells for electrical purposes.

Solar Thermal

Whereas photovoltaic cells process light into electricity directly, solar thermal energy requires using the sun's rays to heat a fluid that is then used to power a steam generator. Some systems use parabolic dishes, which feature a mirror array to concentrate sunlight into a receiver that holds liquid, which can be used to drive a generator.

One example is Solar Two, an innovative attempt by the U.S. Department of Energy and a number of utility companies across the nation. The project was built in the California Desert near the city of Barstow. Solar Two used almost 2,000 sun-tracking mirrors that reflected sunlight onto a receiver on a 300-foot-high tower. Three million pounds of molten salt flow through the receiver, are heated to 1,065°F (400°C), and then are transferred to a hot salt storage tank. When electricity is needed, the super-heated salt is run through a steam engine, which turns a turbine generator. The storage tanks allow the system to continue to generate electricity at night without batteries.

Solar Power from Space

Some scientists think that space is the best place to collect solar energy and so have been casting about for ways to make that possible. In space sunlight is available almost all the time with no clouds or atmospheric conditions to dilute it. In the 1970s NASA studied the idea of using a solar cell array the size of Manhattan in stationary orbit over the earth. The satellite would gather the light and convert it to radio frequency energy. It would then beam that energy to Earth where a receiver would capture it and convert the radio frequency back to electricity.

Japan has invested in such a system and is working to beam solar energy to developing nations a few degrees from the equator from a satellite in low equatorial orbit. A number of tropical nations including Papua New Guinea, Indonesia, Ecuador, and Columbia have agreed to participate in the experiment. Japan hopes to be beaming down energy by the second half of the century.

Wind Energy

From high mountain passes in California to the shores of the North Sea, tall farms of wind turbines stand upright against the blustery force of the winds to capture their force in turbine blades and convert it to electricity. Wind turbines around the world are now producing commercial quantities of electricity without emitting any greenhouse gases.

Wind energy is actually another form of solar energy since wind is driven by temperature and pressure differences on the surface of the planet, which are caused by the sun. Wind turbines—generally with two or three blades—collect the energy of the moving air to drive a generator, which creates electricity. Wind turbines are set on towers in places where winds blow hard and steady. They're different than windmills used on farms to bring up water. These monsters can have blades up to 82 feet (25 meters) long.

Hot Debates

Wind energy is cheap and clean but it has some drawbacks. Though wind farms are generally located in unpopulated areas, local residents complain about the visual pollution these huge wind farms create. Also, migrating birds, particularly hawks and eagles, can get tangled in the turbines and perish. One solution to this problem has been to put wind turbines out to sea.

The U.S. Department of Energy estimates that global winds could provide more than 15 times the world's energy consumption. There is enormous potential for greater use of wind energy in the United States. Midwest states such as South Dakota, North Dakota, Nebraska, and Kansas could become the Middle East of wind power.

In wind-swept northern Europe, wind turbine operators are constructing huge full-scale marine wind parks in the North Sea. There are advantages to offshore windmills besides the lack of visual pollution. Offshore, operators never run out of wind. It blows 90 percent of the time. Wind provides some 28 million Europeans with electricity. About half of them are in Germany, Europe's largest producer of wind-driven energy.

Geothermal Energy

Geothermal energy arises from the earth, usually in the form of volcanoes, hot springs, and steam vents. But there's more down there than that, including decaying substances and friction along fault lines. If we could harness all the energy in just the top six miles of the earth's crust it would equal 50,000 times the energy of all the world's gas and oil reserves.

Steam-dominated sources supply the energy for most geothermal power plants in the United States. The Geysers, near Morgan Springs, California is one of them. Here water descends to relatively shallow depths from 1,000 to 10,000 feet (300 to 3,000 meters), heats up, and returns as hot water or steam. Hot rock could also be used with energy suppliers pumping the water down into the rock and utilizing the heat that returns to run electric generators.

Still the equipment is expensive. At present few major utilities are running any of these operations in the United States. In general, operations are run by small businesses, which turn around and sell the energy to utility companies. Though the demand for geothermal created energy is low in the United States, other countries such as Iceland and New Zealand embrace it enthusiastically.

Climatoids

Though geothermal-created electricity has a high price tag attached to its installation, using geysers or hot spring water to heat homes or businesses is a viable alternative. Geothermal heat pumps can move the water from 5 to 50 feet below the structure up into the building to heat it in the winter and even cool it in the summer, since subsurface temperatures stay nearly constant year round.

Biomass

We get biomass energy from organic materials. The most common biomass material is wood. More than 500 facilities use wood or wood waste to generate electricity. Some use waste materials left over from planting, pruning, and harvesting. We can also make liquid and gas fuels like ethanol and methanol which could be mixed with existing fuels. These would burn relatively cleaner than fossil fuels though they are not as clean as geothermal, wind, or solar.

If we had a fuel crop, it could serve as an interim fuel until cleaner options were more available. One possibility is to use the plant residue from sugarcane, left after milling. Sugarcane was the basis for the world's first renewable biofuel program in Brazil. Corn is presently used to create ethanol in the United States.

Municipal solid waste is composed of 60 percent organic materials and can be used to generate energy. Landfills produce methane gas, which can be captured and used to fire electric generators. Suppliers can use yeast to ferment biomass and convert it to alcohol, which can be used for cars and trucks.

The Ocean as an Energy Source

There is a lot of power in the ocean from thermal, tidal, and wave-generating forces. There have been a number of experiments to capture the sea's reservoir of energy. One experiment in Hawaii and one on the Pacific island of Nauru by the Japanese tried to harness the ocean's thermal energy—using warmer waters at the surface and cooler waters underneath to run a steam turbine. Others have tried to harness the energy of waves, using the surging, the up-and-down, and even the rocking motion of the waves, but none of these experiments have hit pay dirt as yet.

The Canadians and the French have successfully harnessed the energy in the tides. Rising tides enter a reservoir controlled by gates which close at high tide. After the tide goes out, the reservoir releases the water to drive turbines much as if a river were passing through a dam.

Climatoids

Blue Energy of Canada is developing a technology that acts as an underwater wind farm. Turbines are mounted on structures that are set out on the ocean floor in places with strong tides or currents. Water flows through the structures, driving the turbines.

Fuel Cells

Fuel cells are not new. NASA developed the technology as a lightweight power source for U.S. space vehicles, including the space shuttle. Fuel cells work somewhat like a battery. In a battery, energy is stored on electrodes. "The battery runs until the charge runs out, and you have to recharge it," says John D. O'Sulivan, program manager for fuel cell technology at the Electronic Power Research Institute in Menlo Park, California, "but fuel cells provide continuous energy because you just keep adding fuel."

Fuel cells work by allowing oxygen to react with natural gas, methanol, or hydrogen, to produce electricity without combustion. In hydrogen fuel cells, water and heat are the only byproducts.

Both Honda and Toyota have introduced a few fuel cell vehicles in California and more could be on the way. Every major automaker and oil company has a hydrogen or fuel cell research effort under way. Supporters say they recognize that fossil fuels can't last forever. Environmentalists complain that the industry is simply trying to avoid more immediate steps to improve fuel efficiency of conventional automobiles. Though it may be a while, fuel cells have that look of the future.

One of the problems is the price. It costs about $4,000 to make a gas engine. It costs about $40,000 right now to make a fuel cell. Fuel cells have few moving parts so maintenance would be low. They don't consume themselves like batteries, and they remain almost endlessly rechargeable. Producing a similar amount of electricity as one fuel cell from a power plant would dump 1,100 tons of carbon into the atmosphere.

Because hydrogen, the ideal fuel source for fuel cells, is so difficult to create and store, fuel cells that run on gasoline or other less environmentally friendly fuels may precede the pure hydrogen models. Such a vehicle could travel 80 miles on a single gallon. It would emit no pollutants except for CO_2. The increased mileage would cut heavily into that.

A Hydrogen Economy?

Hydrogen has enormous potential. It could just be the prophet that finally leads us out of this sea of fossil fuels. Even now, just adding 5 percent hydrogen to the gasoline-air mixture in internal combustion engines could reduce nitrogen oxide emissions by up to 40 percent. Hydrogen has potential for use as an aircraft fuel. Within a few years hydrogen could be blended with natural gas by electric utilities wishing to reduce emissions. Stand-alone hydrogen-powered fuel cells could provide power for remote and commercial applications.

The major problem with using hydrogen as an energy source is that it doesn't occur naturally in any usable from. Hydrogen almost always bonds with something else and requires energy to break the bond to produce pure hydrogen. This is the rub. It currently takes twice as much energy to create hydrogen than it takes to make electricity.

Climatoids

Engines may be converted to run on hydrogen by the use of a SmartPlug, a device developed by kNew Corporation. SmartPlug replaces spark plugs, combining the functions of fuel injection and spark ignition, and can operate on a number of different fuels, including hydrogen.

Scientists are experimenting with a number of different methods for producing hydrogen. Most involve trying to break good old H_2O into hydrogen and oxygen. One method uses light focused on semiconductors submerged in water. Another uses dissolved metals as a catalyst to create water-splitting reactions. Some types of bacteria work to convert carbon monoxide and water to carbon dioxide and hydrogen. Scientists from the Woods Hole Oceanographic Institution recently boarded a submersible to search the hydrothermal vents at the bottom of the ocean for microbes that may one day be engineered to capture carbon dioxide and to strip hydrogen from methane.

Let's Get Started

So what are we waiting for? The challenge is out there. The dangers of just waiting around and doing nothing are becoming clearer every day. We've been dealing in oil for over a century. It's been good while it lasted, but it's turning on us. Following its path is increasingly dangerous.

It's time to start looking at other options.

The Least You Need to Know

- There are a lot of ways we can start cleaning up the fuels we use right now.
- Solar energy, including spaced-based solar systems, has enormous potential.
- Wind, geothermal, and biomass energy could fill in the gaps.
- Fuel cells run on hydrogen and may someday replace the internal combustion engine.
- If we can figure out a cheap way to create hydrogen, we might have global warming licked.
- Time's running out. Let's get going.

Concerned Organizations

Canadian Rainforest Network (CRN)
Box 2241
Main Post Office
Vancouver, BC, Canada
V6B 3W2
Phone: 604-669-4303
Fax: 604-669-6833

Center for Conservation Biology (CCB)
Department of Biological Sciences
Stanford University
Stanford, CA 94305-5020
Phone: 650-723 5924
Fax: 650-723 5920
www.stanford.edu/group/CCB/

Conservation International (CI)
2501 M Street NW, Suite 200
Washington, DC 20037
Phone: 202-429-5660
Toll-Free: 1-800-429-5660
Fax: 202-887-0193
www.conservation.org

Environmental Defense (ED)
257 Park Avenue South
New York, NY 10010
Phone: 212-505-2100
Toll-Free: 1-800-684-3322
Fax: 212-505-2375
www.edf.org

Greenpeace USA
702 H Street NW
Washington, DC 20001
Phone: 1-800-326-0959
www.greenpeaceusa.org

International Union for the Conservation of Nature (IUCN) or the World Conservation Union (WCU)
Rue Mauverney 28
1196 Gland
Switzerland
Director General's Office
Phone: 41-22-999-0152
Fax: 41-22-999-0015
www.iucn.org

International Tropical Timber Organization (ITTO)
International Organizations Center, 5th Floor
Pacifico, Yokohama
1.1.1, Minato.Mirai, Nishi.ku, Yokohama 220.0012 Japan
Phone: 81-45-223-1110
Fax: 81-45-223-1111
www.itto.or.jp

National Audubon Society (NAS)
1901 Pennsylvania Avenue NW
Washington, DC 20006
Phone: 202-861-2242
Fax: 202-861-4290
www.audubon.org

National Wildlife Federation (NWF)
8925 Leesburg Pike
Vienna, VA 22184
Phone: 703-790-4000
www.nwf.org

Natural Resources Defense Council (NRDC)
40 West Twentieth Street
New York, NY 10011
Phone: 212-727-2700
Fax: 212-727-1773
www.nrdc.org

The Nature Conservatory (TNC)
4245 North Fairfax Drive, Suite 100
Arlington, VA 22203-1606
Phone: 1-800-628-6860
www.tnc.org

Rainforest Alliance (RA)
65 Bleeker Street
New York, NY 10012

Phone: 212-677-1900
Toll-Free: 888-MY-EARTH
Fax: 212-941-4986
www.rainforest-alliance.org

Sierra Club
85 Second Street
San Francisco, CA 94105
Phone: 415-977-5500
Fax: 415-977-5799
www.sierraclub.org

Smithsonian Institute (SI)
Washington, DC 20560
Phone: 202-357-2627
Fax: 202-786-2377
www.si.edu

Survival International (SI)
11-15 Emerald Street
London WC1N 3QL
United Kingdom
Phone: 020-7242-1441
Fax: 020-7242-1771
www.survival.org.uk

Union of Concerned Scientists (UCS)
2 Brattle Square
Cambridge, MA 02238
Phone: 617-547-5552
Fax: 617-864-9405
www.ucsusa.org

United Nations Environment Programme (UNEP)
United Nations Avenue, Gigiri
P.O. Box 30552
Nairobi, Kenya
Phone: 254-2-621234
Fax: 254-2-624489/90
www.unep.org

The Wilderness Society (TWS)
1615 M Street NW
Washington, DC 20036
Phone: 1-800-843-9453
www.wilderness.org

World Wildlife Fund for Nature (WWF)
Avenue du Mont-Blanc
CH-1196, Gland
Switzerland
Phone: 41-22-364-91-11
Fax: 41-22-364-53-58
www.panda.org

World Forestry Center (WFC)
4033 SW Canyon Road
Portland, OR 97221
Phone: 503-228-1367
Fax: 503-228-4608
www.worldforestry.org

Worldwatch Institute (WI)
1776 Massachusetts Avenue NW
Washington, DC 20036
Phone: 202-452-1999
Fax: 202-296-7365
www.worldwatch.org

Further Reading

Books on Climate and Climate Change

Alley, Richard B. *The Two Mile Time Machine: Ice Cores, Abrupt Climate Change, and Our Future*. Princeton University Press, 2000.

Casten, Thomas R. *Turning Off the Heat*. Prometheus Books, 1998.

Committee on the Science of Climate Change, National Research Council. *Climate Change Science: An Analysis of Some Key Questions*. National Academy Press, 2001.

Goldstein, Dr. Mel. *The Complete Idiot's Guide to Weather*. Alpha Books, 1999.

Harris, Paul G., ed. *Climate Change and American Foreign Policy*. Hampshire, UK: Palgrave, 2000.

Houghton, John T. *Global Warming: The Complete Briefing*. Cambridge University Press, 1997.

Lamb, H. H. *Climate, History and the Modern World*. Routledge, 1997.

Lutgens, Frederick K. *The Atmosphere: An Introduction to Meteorology*. Prentice Hall, 1989.

National Research Council. *Abrupt Climate Change: Inevitable Surprises*. National Academy Press, 2002.

Revkin, Andrew. *Global Warming: Understanding the Forecast*. Abbeville Press, 1992.

Root, Terry and Stephen H. Schneider, ed. *Wildlife Responses to Climate Changes*. Island Press, 2002.

Somerville, Richard C. J. *The Forgiving Air: Understanding Environmental Change.* University of California Press, 1996.

Assessment Reports of the Intergovernmental Panel on Climate Change ("IPCC")

Climate Change 2001: *Impacts, Adaptation and Vulnerability.* Cambridge University Press, 2001.

Climate Change 2001: *Mitigation.* Cambridge University Press, 2001.

Climate Change 2001: *The Scientific Basis.* Cambridge University Press, 2001.

Field Guides

Chartrand, Mark R. *National Audubon Society Field Guide to the Night Sky.* Knopf and Chanticleer Press, 1997.

Day, John A., and Vincent J. Schaefer. *A Field Guide to the Atmosphere.* Houghton Mifflin, 1999.

Ludlum, David M. *National Audubon Society Field Guide to North American Weather.* Knopf, 1991.

Books on the Environment

Berger, John J. *Charging Ahead: The Business of Renewable Energy and What It Means for America.* University of California Press, 1998.

Burton, John, ed. *The Atlas of Endangered Species.* Macmillan, 1998.

Ehrlich, Paul R., and Anne H. Ehrlich. *Healing the Planet.* New York: Addison-Wesley, 1991.

Gay, Kathlyn. *Rainforests of the World.* ABC-CLIO, Inc., 2001.

Harrison, Paul. *The Third Revolution: Environment, Population, and a Sustainable World.* New York: St. Martin's Press, 1992.

Hurst, Phillip. *Rainforest Politics: Ecological Destruction in Southeast Asia.* Humanities Press International, 1990.

Myers, Norman. *The Primary Source: Tropical Forests and Our Future.* W. W. Norton & Company, 1992.

Revkin, Andrew. *The Burning Season: The Murder of Chico Mendes and the Fight for the Amazon Rainforest.* Houghton Mifflin, 1990.

Wilson, Edward O. *The Diversity of Life.* Harvard University Press, 1992.

Glossary

acid rain   Rain, snow, sleet, hail, and fog, which because of the addition of pollutants becomes more acidic than normal.

aerobic   Living or occurring with oxygen.

albedo   The ability of a surface to reflect back the light that falls upon it.

anaerobic   Living or occurring without oxygen.

anemia   A disease in which body tissues are deprived of oxygen by a reduction in the number of red blood cells or inadequate amounts of an essential protein called hemoglobin.

auroras   The aurora borealis and the aurora australis, known as the northern and southern lights, are vivid nighttime displays of colored light created when charged particles from solar flares strike the earth's atmosphere near the magnetic poles.

barometer   The gauge used to measure the pressure of the atmosphere.

bartonelosis   A bacterial disease transmitted by sand fleas that causes wart-like sores on the skin in milder cases and anemia in more severe cases.

Big Bang   The theory that the universe we know began its existence about 13 billion years ago as the explosion of a small super-dense concentration of matter.

biodiversity   Describes the number of species in a given environment. It is shorthand for "biological diversity."

biological pump   The biological pump is a natural mechanism whereby carbon from the atmosphere and surface waters of lakes and oceans are cycled into deep waters and sediments. Plants take up the carbon dioxide at the surface of the water. Marine organisms then take up the plants and deposit a fraction of the carbon dioxide on the bottom as waste.

blizzards   A term that was used to describe the dust storms that swept the south central plains of the United States during the dust bowl era of the 1930s.

boreal forest   The northernmost type of forest dominated by coniferous trees such as the evergreen spruce, fir, and pine, and the deciduous larch or tamarack.

broadleaf   Trees with broad, flat leaves that are often deciduous—they shed their leaves each year.

carbon burial   The process whereby silicate rocks erode and the chemical breakdown of the rocks converts carbon dioxide to bicarbonate, which is washed to the oceans where it ends up as carbonate sediments at the bottom of the sea.

carbon dioxide   A colorless, odorless gas that passes out of the lungs during respiration. It is also given off by volcanoes. It is the primary greenhouse gas, because it is long lived (approximately 100 years) and causes the greatest amount of global warming of the long-lived gases.

carbon sinks   Areas such as forests, grasslands, and the oceans, which naturally take up carbon dioxide from the atmosphere.

celestial sphere   A huge sphere or dome that ancient observers believed surrounded the earth, upon which the stars were fixed. Today, astronomers use the idea as a way to map the location of stars relative to observers on earth.

cirrus clouds   Clouds that are high and wispy.

clear cutting   Clear cutting is the practice of removing all the timber in a given area, rather than going in and selectively cutting trees. Foresters claim it is necessary for light to rejuvenate seedlings. Environmentalists claim it is a blight, often driven more by economics (it's cheaper to cut down all the trees) than good forestry practices.

climate   The term that refers to the average and common weather conditions of an area over a number of years.

climate change   The natural and manmade phenomena that have made the earth's atmosphere warmer or colder over the ages.

condensation nuclei   The microscopic particles in the air around which cloud droplets and raindrops form.

conifers   Trees with needle-shaped leaves that are evergreen—they stay green all year.

converging currents   Currents in the atmosphere that occur where winds meet or where more air enters an area than departs.

Coriolis effect   The effect caused by the spherical shape of the earth which results in winds veering to the right in the northern hemisphere and to the left in the southern hemisphere.

critical density   That point in the evolution of a star where gas and dust come together close enough for gravity to take over and bind the elements together.

cumulus clouds   Clouds that are puffy and unthreatening; or gathering, rising, and thunderous.

diverging currents   Currents that occur when winds depart in different directions or when more air leaves than arrives.

dust bowl   An area in the south central plains of the United States that was devastated by drought in the 1930s.

eccentricity   The shape of the earth's orbit which varies from round to elliptical.

ecotourism   Environmentally friendly tourism. In North America it refers mostly to minimizing visitor impact. In third world nations it more often refers to supporting local economic alternatives to depleting natural habitats.

ectotherm   An organism that regulates its body temperature largely by exchanging heat with its surroundings.

endemic   Those species of plants or animals that are found only in restricted areas—a single island, a single river or lake, a single mountain range, or a single mountaintop.

Energy Star   A label on consumer products sanctioned by the U.S. Environmental Protection Agency to help identify energy-efficient products.

ethnobotanists   Scientists who study the cultural and biological relationship of native peoples and the plants in their environment.

frontal cyclones   Large spinning weather systems that occur along turbulent fronts.

frost fairs   Recreational and social events held on the Thames River from the beginning of the 1600s to the beginning of the 1800s during the Little Ice Age, when the river froze over frequently. Since then, the world has gotten warmer, and now the Thames seldom freezes.

gas giants   These are the huge gaseous planets Jupiter, Saturn, Uranus, and Neptune, which have enormous thick atmospheres.

glaciers   Moving masses of ice and snow that accumulate in areas where the rate of snowfall exceeds the rate at which the snow melts.

global warming   The term that refers to recent increases in global temperatures caused by man.

greenhouse effect   This is the consequence that certain gases, chief amongst them carbon dioxide, have on the earth's climate. Though they constitute less than 1 percent of the atmosphere, they act like a blanket covering the earth. Without the greenhouse effect, the average global temperature would be around 0°F instead of the current toasty 59°F.

Holocene   The current or most recent epoch of geologic time, ranging from the present back to the time (about 11,000 years ago) of almost complete withdrawal of the glaciers.

hot air   A term used by environmentalists to apply to greenhouse gas reductions that resulted from weakened economies and the resultant reduction in manufacturing rather than actual efforts put into making those industries operate cleaner and more efficiently.

hurricanes   Violent storms that originate in the tropics and whose winds exceed 73 mph.

hybrid   A vehicle that utilizes both an electrical motor and a gas-powered engine which work in tandem to decrease fuel consumption.

hydroelectric   Power or electricity created by the force of water driving a generator.

ice age   Periods of geological time where Earth was covered by substantial portions of glacial ice. The glaciation of the Pleistocene, which began more than 1.8 million years ago, was not continuous but consisted of several glacial advances interrupted by interglacial stages. Many scientists believe we are currently in such an interglacial stage, which began about 11,000 years ago.

ice sheets   Huge glaciers like those that cover Greenland and Antarctica.

inclination of the axis   A measure of the number of degrees that the earth's axis is tilted away from the plane of its orbit around the sun.

industrial revolution   The period in the economic development of a nation when handicraft production gives way to machine production.

initial conditions   The starting point of a weather forecast that takes into consideration temperature, wind speed and direction, air pressure, and humidity measured from points in a grid overlaying the landscape.

interglacial   The period of relatively stable climate between ice ages. The earth is currently in an interglacial period.

isotopes   The varying forms of an element that have closely related properties but different atomic weights.

jet streams   Strong ribbons of horizontal winds that are found about six to ten miles above the ground in the area between the troposphere, the lower layer of the atmosphere, and the stratosphere above it.

kivas   The ceremonial chambers that surrounded the towns and cliff dwellings of the Anasazi Indians. Their structure was influenced by an earlier day when these Native Americans built homes of covered pits.

mesosphere   The layer of the atmosphere that lies on top of the stratosphere and runs upward to about 50 miles (80 kilometers).

methane   Also called swamp gas, it is the principal component of natural gas. It's given off as part of the decomposition that occurs in wetlands, swamps, and bogs, as well as landfills, rice paddies, and the guts of cattle. It's more efficient at trapping heat than carbon dioxide, but occurs in small concentrations in the atmosphere and has a lifetime of only about 10 to 12 years.

nimbus clouds   Rain-bearing clouds.

obliquity   The term that refers to changes in the tilt of the earth's axis.

old growth forest   The type of forest that has not been previously logged. It represents a more natural and diversified mix of tree and plant species.

ozone   The gaseous, almost colorless form of oxygen that protects the earth against ultraviolet radiation in the upper atmosphere, but is part of the chemical pollution nearer to Earth.

permafrost   The vast stretch of permanently frozen subsoil that stretches across the extreme northern latitudes of North America, Europe, and Asia. This land, too cold for the growth of trees, generally marks the northern limits of the forest.

photosynthesis   The process whereby green plants take in carbon dioxide, mostly through their leaves, and water, mostly through their roots, and mix them together in the presence of sunlight and chlorophyll to produce glucose, a carbohydrate utilized by the plant as energy.

phytoplankton   The microscopic plant life found drifting in the upper sunlit portion of the sea or freshwater, which are the basis of the food chain for other marine organisms.

plate tectonics   Plate tectonics is the theory that the earth is made up of plates that move around, bump, and collide with each other in ways that create some of the earth's most spectacular land forms. The Himalayan Mountains were created when the floating landmass that was India plowed into southern Asia.

precession   The term that refers to the amount of wobble in the earth's axis.

prominences   Huge loops of gases that rise from the surface of the sun and extend far out into space.

proxies   Methods of determining values such as temperatures and rainfall by using substitutes, which give indirect measurements.

relative humidity   The measure of the air's capacity to hold moisture versus the actual moisture present. To get rain you have to have 100 percent humidity.

Renaissance   A term that refers to the great revival of art, literature, and learning that occurred in Europe from the 1300s through the 1500s.

sediments   Fragments of inorganic and organic material that are carried and deposited in layers by wind, water, or ice.

smog   A kind of air pollution that comes in two types. The gray smog of older industrial cities like London and New York derives from the massive combustion of coal and fuel oil in or near the city. The brown smog characteristic of Los Angeles and Denver in the late twentieth century comes from automobiles.

solar flares   Violent eruptions of matter and energy on the sun that spew light and charged atomic particles into space.

storm surges   The tremendous rises in tide that storms like hurricanes push ahead of them.

stratosphere   The layer of the atmosphere just above the troposphere. It rises from 7.5 miles (12 kilometers) to an average of 31 miles (50 kilometers).

stratus clouds   Clouds that are flat and low to the ground.

summer solstice   The time period when the north pole is inclined 23.5 degrees *toward* the sun. In the northern hemisphere this falls on June 21.

systems   From an ecological point of view, systems are niches in the environment that incorporate not just one but various elements such as plants, animals, water, and seasonal weather.

terrestrial planets   The smaller rocky planets Earth, Mars, Mercury, and Venus, which may or may not have thin atmospheres.

thermocline   The boundary between the ocean's warm and cold waters.

Thermohaline Circulation (THC)   A global ocean current that operates like a conveyor belt running through shallow and deep-sea waters to connect the major oceans of the planet.

thermosphere   The uppermost layer of the atmosphere, rising from 50 miles (80 kilometers) to around 400 miles (644 kilometers), though it really has no well-defined upper limit.

tillite   A layer of rock and clay that are evidence of glacial deposits.

troposphere   The bottom layer of the atmosphere, rising from sea level up to an average of about 7.5 miles (12 kilometers).

upwellings   The upward movement of deeper water to the surface.

urban heat island effect   A term used to describe the fact that urban environments are usually warmer than surrounding rural areas.

weather   The current conditions of temperature, humidity, wind, and precipitation.

winter solstice   The time period when the north pole is inclined 23.5 degrees *away* from the sun. In the northern hemisphere this falls on December 21.

Younger Dryas event   The event occurred about 12,800 years ago during a period of increasing warmth, when the world was suddenly and very drastically thrown back into the last glacial period of the ice age. It remained that way until about 11,500 years ago when it swung back out.

Index